my tiny
home farm

Francine Raymond
Photography by Bill Mason

my tiny home farm

SIMPLE IDEAS FOR SMALL SPACES

PAVILION

First published in the United Kingdom in 2017 by
Pavilion
43 Great Ormond Street
London
WC1N 3HZ

ISBN 978-1-91090-472-5

A CIP catalogue record for this book is available from the British Library.

10 9 8 7 6 5 4 3 2 1

Reproduction by Mission Productions Ltd, Hong Kong
Printed and bound by 1010 Printing International Ltd, China

This book can be ordered direct from the publisher at www.pavilionbooks.com

CONTENTS

INTRODUCTION

Smallholders come in all shapes and sizes. Some just grow fruit and veg and keep a few chickens and bees, while others go the whole hog. The luckiest get to maximize the symbiotic relationship between land and animals, producing wholesome food for their families by running mixed holdings where livestock grazes fertile land.

In an ideal world, we would all have access to healthy home-grown produce, and although most of us will never have space to grow a whole larder, we can all make our lives a little richer by planting a few seasonal delights, maybe offering a home to a couple of bantams, and making life easier for pollinating insects wherever we live.

With land at a premium, how can we start to enjoy the benefits of the good life? The answer is to buy from local markets and encourage local start-ups, to patronize cafés and restaurants that use ingredients grown nearby and to spread the green word whenever we can, especially to the next generation.

If you live somewhere with no outside space at all, seek out your local community garden and join in; find out if there's a community orchard nearby and volunteer. You'll share the harvests, enjoy the growing experience and make new friends. Even a spot of guerrilla gardening might give you a tiny plot, a verge or tree pit in which to grow fruit, veg and herbs.

If you have no time for regular commitment, try a farm holiday. Organizations such as Workaway offer experience on farms all over the world where you'll have fun and learn at the same time. Seasonal fruit picking is an option, too. If you have cash to spare and want to put your feet up, stay in a hotel that trumpets its home-grown menu, home-raised meat and green credentials.

Allotments offer city-dwellers space to grow produce, and a few allow poultry and beekeeping too. Swedish koloni, German Schrebergärten, French jardins familiaux, Italian orti sociali, Dutch volkstuinen, Russian dachas and US community gardens all provide growers with plots at low rent, but there are usually long waiting lists and strict regulations often apply.

Those with balconies, backyards and flat roofs can grow surprising amounts of edible produce by maximizing space with containers, climbers and vertical gardens, though it's important to remember to water and feed them regularly. Rooftop beekeeping is successful in cities all over the world, with bees foraging in parks and gardens within a 8km (5 mile) radius of their hives.

With a small front or back garden, the range of edible goodies you can grow increases mouth-wateringly. Trained fruit trees on a dwarf rootstock provide a delicious range of fruit, while raised beds and containers full of vegetables and herbs rotated with care and sown in succession will feed a small family on salads and other seasonal veg. If you have room for a compact greenhouse, or even just a cold frame, you can lengthen your growing season.

With a larger garden, poultry keeping becomes a possibility, provided Mr Urban Fox or other predators aren't visitors. With all creatures, however, giving them enough space is of paramount importance since they won't thrive without it, so try bantams or call ducks and just keep a trio. An 0.4ha (1 acre) garden could be home to a couple of pigs, bought as weaners then grown to slaughter-size in autumn.

Animal welfare should be uppermost in the smallholder's mind; if you are thinking of keeping stock of any kind you must go on a course first and learn exactly how to look after your animals. Wherever you live, keeping stock will be heavily regulated, for both the animals' sake and the safety of the general public, so make sure you are aware of the legal hoops you'll need to jump through.

If you are in the lucky position to be buying land for a mixed smallholding, then this is your wish list:

- Ground that isn't wet, low-lying or prone to flooding.
- Soil that's fertile with good grazing.
- A nearby abattoir with a local vet who deals with your type of livestock.
- Neighbours likely to be sympathetic to your farming plans.
- Easy access for farm vehicles with no public rights of way.
- Existing standpipes, farm buildings and fencing would be the cherry on the cake.

Sometimes a minus can be turned into a plus: wetland could grow watercress or be used for farming fish; a sandy, light soil could make a comfy home for pigs; and a south-facing slope might make an award-winning vineyard. But whatever you grow, the land needs to be weed- and pest-free and fertile.

After caring for your livestock, looking after your soil should be your next priority: improve it with your own compost and enrich it with your own fertilizer (green or animal manure), raising productivity and dealing with waste products. Provide habitat and food for wildlife, and the wildlife will help to control pest depredations and pollinate your crops.

Save cash by gathering seeds, building DIY structures and collecting firewood for fuel, and use your ingenuity to make the most of your crops at harvest time with tried and tested preserving techniques – useful skills whatever the size of your plot. With surpluses, you'll need marketing skills and an understanding of the health and safety regulations to sell your produce online or at farmers' markets.

I've loved getting to know the smallholders I visited in the making of this book, and I hope their experiences will help you to take the first steps to a more self-reliant, healthy and sustainable life, producing food that you and your family love to eat.

GROWING
ALL OVER
THE WORLD

The world is full of wonderful produce and we can all learn from one another. Changes in national diets have been introduced by pioneers and newcomers alike, from Francis Drake and his potatoes from the New World to holidaymakers bringing back souvenirs of wonderful meals and immigrant communities bringing their cuisines to their adopted homes.

Seed companies have responded to new tastes and the Internet makes it possible to buy internationally. Co-workers on allotments have influenced change in what's grown in veg plots and back gardens, and television programmes, cookery books and magazines publicize each fresh find. Developments in animal breeding and husbandry sometimes improve yields, while success in preserving ancient and rare breeds maintains useful traits for future generations.

So all the world's ingredients can be on our home-grown menu – all we need is the warmth of the sun (with maybe a little help under glass), fertile soil with adequate water, and the desire to try something new.

*Working together in the fresh air
cements relationships and sharing
know-how produces great results.*

VEG UNITED NATIONS

**At first glance, allotments seem unchangeable, part of
a way of life that has been the same for centuries. But look
closely, and you'll see fruit, veg and herbs unrecognizable
to gardeners 50 years ago, almost usurping the potatoes,
carrots and turnips that were once maincrop. Traditional
produce is being supplanted by all sorts of delicious
ingredients that are changing the face of our national menus.**

And check out the allotmenteers, too: gardeners from all over
the world, enjoying growing together. I visited allotments on
sites in towns, outside villages and right in the middle of cities;
spaces set aside by local authorities, mostly a century or more
ago, so that families had space to grow their own food.

These communal gardens have flourished historically during times of need, depression and war. They have fallen in and out of fashion, but are now happily in demand again, with long waiting lists and families sharing plots. Not only are they used to grow food, they are places to relax healthily in the open air and get to know your neighbours.

I met Gerardo and Carmella Crolla on their English hillside plot one Sunday morning, interrupting their artichoke harvest. Carmella picks them young to stuff with meat, herbs and breadcrumbs and serve with a tomato sauce, and then any surplus is frozen or preserved in oil (sott'olio) or in vinegar (sott'aceto). I noticed there were artichoke plants of the 'Romanesco' variety growing all over the place: generous hand-outs from Gerardo to his neighbours, though Carmella says, 'They leave them growing too long, they'll be like leather!'

In neat rows, there are borlotti beans to dry for winter stews; scrambling zucchini and squash plants; chicory to blanch and eat in calzone; tomato 'San Marzano' for sugo; and cima di rape to eat with pasta orecchiette. In the cold frame, I saw agretti, a plantain-like salad to eat young, and basil and rocket – only on sophisticated menus until ten years ago, but now everywhere. In return, Gerardo is growing a clump of rhubarb given to him by his friend on the next plot – a delicacy unheard of in Italy.

Gerardo, Carmella and their grandson stand proudly among their produce: rows of vegetables to serve at family meals and a tiny vineyard for table grapes.

The Crollas grow all their own veg, plus plums, apples, peaches and strawberries. Eight-year-old grandson Gerardo junior, visiting from Wales, says he loves his nonna's cooking, but she complains her daughters don't have time to grow or cook. Looking around, not many of the younger generation are in evidence on this allotment either.

Gerardo collects and dries his own seeds, and often brings huge, cheap packets back from his annual visits to Italy, complaining that seeds here are mean and don't germinate well. But at least these unusual varieties are available now for everyone to try. Seed companies have responded to demand, and we're all the richer.

The skies may be grey, but it is possible to grow Caribbean, Chinese, African and Pakistani ingredients here too. Stunning scarlet Jamaican callaloo or leaf amaranth (whose leaves and stems are eaten cooked like spinach), and scorching hot Scotch bonnet chilli peppers, pumpkins and broadleaf thyme all need a little early heat. I met a Pakistani gardener growing sweet pumpkins, coriander and fennel in her tiny greenhouse.

Chinese salads such as pak choi, mizuna and mibuna are popular with allotmenteers and even available in supermarkets, thanks to vegetable evangelist Joy Larkcom, but less common tatsoi and wong bok are easy late-summer salad leaves too. Okra, used to thicken stews, and aubergines are commonplace in urban allotments, but one day we may also see highly nutritious African jute mallow (*Corchorus olitorius*), spider plant (*Cleome gynandra*) and slenderleaf (*Crotalaria brevidens*) all growing happily alongside onions, potatoes and carrots.

HOW TO:
PLANT A POTATO BUCKET

Potatoes take up a lot of space, which may be at a premium in a small veg garden. Nothing beats the taste of home-grown new potatoes, though, so why not plant a few in a bucket or old dustbin? It's fun to do, especially for children, and may avoid some of the more common soil-borne diseases that may come from planting them in a border.

Each potato needs about 10 litres (just over 2 gallons) of compost to grow in, so a bin will hold about four plants, and a small bucket just one. Salad potatoes or first and second earlies are best for this method of cultivation.

1. Place your potatoes in a cardboard egg tray, standing them rose end up (the blunter end, with 'eyes'). Leave them in a light, cool place until they start to shoot.

2. Place some crocks in the base of the bucket for drainage. Quarter-fill it with peat-free multipurpose compost, then pop in your spuds, sprouts uppermost.

3. Add another quarter of compost and keep adding more as the green shoots grow until you reach the rim of your container. Give liquid feed fortnightly and remember to water regularly or you will harvest potatoes the size of marbles. Harvest your potatoes as soon as the foliage begins to turn yellow and die down.

Keeping hens is a good introduction to animal husbandry, possible in even the tiniest garden.

FROM BANKER TO SMALLHOLDER

Formerly a banker, Hans Meijer is now retired to the Dutch province of Drenthe, where he works as a sculptor and keeps chickens on his smallholding. He is also treasurer of the Dutch Hobby and Smallholder Association. He keeps his birds in old stables in small flocks of 6–8, each presided over by a fine cockerel, and has been breeding chickens since 1984 – though he started raising young cockerels for the table at the age of only 12.

Drenthe has its own breed of hen, the Drentshoen, but Hans keeps another rare breed: the Hagheweyder, developed in the 1980s especially for smallholders as a dual-purpose layer/table bird. In the UK, the Ixworth was bred for meat and eggs, while in France, the Poulet de Bresse is the table bird par excellence, judged the best by chefs through the centuries from Brillat-Savarin to Heston Blumenthal. These hens are raised only by a selected group of breeders, live a free-range life of luxury in spacious surroundings for 16–20 weeks, and sell for premium prices.

Most commercial table birds are bred under patent and sold in large quantities, making it difficult for smallholders to obtain them. However, these producers tend to prefer not to buy creatures designed to reach slaughter weight at just six weeks old, with little flavour. Most smallholders keep crossbreeds that make for better eating with a longer life, using their excess cockerels for the table and keeping the hens for egg-laying.

Hans keeps his mixed flocks of layers and table birds in old stables and lets them roam safely in large pens. You don't need a cockerel for your hens to lay eggs.

Access to fresh water that's conveniently sited is crucial for all smallholdings.

Crosses that lay well and are good to eat include Light Sussex/Wyandotte, Rhode Island Red/Indian Game and Dorking/Indian Game. There are other breeds specially designed for the table including Jersey Giants, Delaware and Buckeye.

Poultry-keepers who hatch eggs soon realize that at least half their output will be male. Kept separately from the age of 12 weeks (when you can tell what sex they are) to ten months when mature, these cockerels will make good eating. As Hans says, 'We have an obligation to treat animals well, to give them a good life, to butcher them with respect and be grateful for the meat we eat. The first few times it's very difficult, but it soon becomes a routine – though one I carry out gently with care, holding the bird securely and dislocating the neck quickly.'

It's very important to learn to cull birds by watching, so go on a course or get an expert to teach you and help you with the first few attempts until you are calm and confident; a botched job is unpleasant for all concerned. In most countries you can slaughter birds for your own consumption, but check with the authorities first. The Humane Slaughter Association and the Poultry Club give an insight into what's involved.

- Starve birds for 12 hours prior to slaughter, but give access to water.
- Do the deed out of sight and sound of the rest of the flock; during the evening when birds are sleepy is the best time.
- Dislocation of the neck renders the bird unconscious before death, but in many countries electrical stunning must be performed first.
- Leave the carcass hanging upside down to pool the blood into the neck.
- Pluck the feathers while the carcass is still warm.
- Hang the carcass in a hygienic place to tenderize the meat for 2–7 days depending on the temperature, then butcher, gut and truss.

Hans told me, 'The meat of our chickens tastes good. They are well fed with good feed and greenery, they have plenty of space and are killed without unnecessary stress, all of which is translated into the taste of the meat.'

*Don't miss out on growing
space on balconies,
terraces and rooftops.*

UP ON THE ROOF

Artist Pamela Reed is also known as Brooklyn Farm Girl. She describes herself as a food blogger, recipe maker, urban gardener and cat mother of five. In 2005, she and husband Matthew moved to the capital from art school in Pittsburgh. Every summer, they missed the Pennsylvania gardens in which they had been brought up, and with no soil and no backyard, they weren't daunted – they just set up a few containers six storeys up on their roof and got gardening.

They started small, but now it's a mini farm. Broccoli and cauliflower, tomatoes and peppers, soya beans for edamame, sugar snap peas, potatoes, onions, carrots, salads and kale – all flourish in pots, bags, beds and boxes, while pumpkins and watermelons scramble up posts. They are protected by fleece on frames to combat the cold winds and the couple check the weather forecast and wind speeds on their phones. They lost everything when Hurricane Sandy struck in 2012, but last season their rooftop garden yielded an amazing 208kg (458lb) of vegetables.

All their food leftovers are composted and mixed with cocoa bean waste to make a lightweight soil and they collect as much rainwater as they can. For the first year they carried all their water up flights of stairs, but now they've hooked hoses together from an adjacent washroom and have water on tap.

Pamela's plants are grown under glass from seed in modules under grow lights.

The couple tried to keep hens. Pamela loved them and fed them leftover vegetable treats, but unfortunately the keeping of hens breached the building regulations and they had to go, though a growing number of New Yorkers are keeping hens in backyards and community gardens. But bees are definitely on Pamela's agenda, as beekeeping is a popular hobby in New York City, which has its own beekeepers' association.

For economy's sake, all Pamela and Matthew's plants are grown from seed in an indoor shelving system with grow lights. Plantlets are not potted out until the temperature is safely frost-free. Pamela says: 'Everything tastes so much better full of love.' She blogs to encourage us all to grow food, no matter how small or inhospitable our growing space, and then concocts mouth-watering recipes for online visitors to try out at home. Her favourite day ends with the couple taking their grill up on the roof, and cooking supper right there up among the veg beds.

Pamela's recipes are simple and straightforward, making the most of fresh produce picked straight from the plot. Gluts are dealt with quickly and sensibly; her recipe for roasted creamy cherry tomato sauce blitzes tomatoes roasted with olive oil, garlic and basil into meal-sized portions of sauce for pasta or pizza, for example.

There's no limit to what can be grown, from vine fruits to root crops and salads.

Lightweight compost and containers, plus good staking against the prevailing winds, are essential for those wanting to grow food up among the clouds.

Pamela was recently invited by Michelle Obama to see the White House vegetable garden that supplies food for the family and staff. The First Lady started an initiative called Let's Move to encourage people to take exercise and eat healthily, with the aim of inspiring all Americans to grow and cook their own food. As Pamela says, 'If we all cooked at home, think of the impact it would have on overall health, and imagine the joy and satisfaction we'd all get from presenting a meal that's not just home-cooked but home-grown too.'

Pamela and Matthew's farming ambitions are slowly outgrowing their space, so they're embarking on a new adventure: buying land and dreaming of a large garden for fruit and vegetables, a pond and lots of animals, 'and a modern shipping container home with a view of the Catskills'. Follow their upstate adventure at brooklynfarmgirl.com. With their energy and skill, it won't be long before that dream comes true.

MOBILE IN THE CITY

The middle of a 27ha (67 acre) building site in the Kings Cross development area in London seems an unlikely place to find apple trees growing, chillies and tomatoes ripening, bees buzzing, hens laying and people enjoying eating locally grown produce. But there it is – or there it isn't. This is a mobile garden where everything is grown in containers, routinely moved as the area develops. At the moment it's in its fourth location, next door to the natural outdoor swimming pond.

This project is run by Global Generation, a charity that gives young people from local London boroughs the opportunity to learn how to create a sustainable future. It connects them to the natural world with hands-on training after school from local businesses and volunteers who teach leadership, enterprise and communication skills and aim to bring a sense of fun and adventure to the city, as well as planting seeds to encourage the smallholders of the future.

When space and funds are at a premium, any container will do to grow crops – and if they're recycled, that's even better.

Using on-site building materials, from skips to builders' bags, the students have fashioned everything they need on the cheap. Workstations and facilities have been made from surplus building materials, and will be transported to the next site when the café and garden move on.

With help from architecture students, seven structures have been built from recycled on-site building materials, giving plenty of inspiration to smallholders who want eco-friendly buildings on the cheap. Skips have been turned into vegetable beds; planters made from scaffold boards and builders' bags have appeared all over the site; reclaimed sash windows have been turned into vertical greenhouses, railway sleepers piled with earth-filled coffee sacks have become work spaces: polytunnels made of water-pipes and polythene help to shelter plants, while home-made comfrey juice fertilizer, wormery compost and rainwater harvesters make plants grow big and healthy.

The garden is managed by Paul Richens, who describes himself as a worm-botherer since the age of three, and has grown organic vegetables in small city spaces ever since. He says, 'This isn't a garden that will last forever – it's only temporary, but we always travel hopefully.' From

the garden, produce goes straight to the Skip Garden Kitchen where it's cooked and served to the public. The ever-changing menu includes seasonal salads with flowers, herbs, veg and fruit juices, and cakes such as tempting chocolate courgette or lemon thyme polenta.

Hens are kept in a plastic Eglu henhouse inside a polyhedron run made of bamboo built round a central salvaged birch tree trunk, designed by UCL student Valeria Vyvial. It houses a small flock of hybrid hens that lay every day and provide plenty of eggs for the café. The walls are patterned to look like falling leaves and perforated to let in light and air.

Where hens are housed in a smallish run, this will need to be moved regularly to avoid a build-up of parasites in the soil. Covering the surface of the run with hardwood chips will keep the flock safe, especially if the chips are regularly raked and washed with a garden hose, then treated with poultry disinfectant and sanitizing powder. Keep all chicken feed in a lidded metal dustbin and remember to clear away any leftover food at night so that you don't attract unwelcome diners.

The Skip Garden is a magnet for local bees, thanks to the rich diversity of plants. The garden hosts workshops on how to keep bees in urban spaces – an essential learning process, along with a mentor, if you want to be a beekeeper (and of course you need to be sure that you're not allergic to bee stings). There's no law that says you have to let neighbours know you're intending to keep bees, but it's polite to do so. The charity Urban Bees offer to put those who want to keep bees in touch with others who have the space. Let's make life easier for bees, and grow more native plants that will flower throughout the year – specifically those with single flowers, which bees prefer rather than complicated double flowers where the nectar and pollen are harder to access.

The gardeners on these Swedish koloni plots grow salads in raised beds, herbs and decorative plants in hanging pots and potatoes in bins.

SWEDISH KOLONI GARDENS

Scandinavians have access to some of the most unspoiled countryside in Europe and love making the most of it, but with only a short growing season, they are not renowned for their interest in productive gardening. Fruit trees surrounded by lawns are the norm. This wasn't always the case: tucked away on the outskirts of the major cities are remnants of what used to be a thriving allotment culture.

Eija Niskakavi is one of a new breed of Swedish gardeners who are rediscovering the joys of koloni trädgårdar, or allotments. She has bought a stuga – a tiny cottage – with a small strip of land, densely packed in next to other plots just outside Gothenburg. These stugor are not for year-round living, because water is only available during summer months.

Koloni gardens were created in the late 19th century by philanthropic employers for the urban poor who had moved to cities as a result of the industrial revolution. In Sweden, as in most parts of Europe, the population had been mainly rural, living off the land, but as numbers doubled the land was no longer able to sustain its inhabitants, and the mechanization of labour put further pressures on the poor.

Inspired by allotments in Copenhagen, Social Democrat politician Anna Lindhagen (1870–1941) encouraged the setting aside of small parcels of land for town-dwellers so that they could improve their meagre existence with home-grown vegetables and fruit, and enjoy spending time outside in the fresh air. Over the years these spaces developed into holiday homes with pretty gardens, where gardeners chatted over picket fences and children played.

Although the stugor continued to be popular, only a few die-hards carried on caring for their plots – in the same way that allotments fell into disrepair in the UK in the period after the Second World War, when producing food wasn't a priority. It took the return to 'the good life' in the 1970s for growing your own to become fashionable again. Nowadays, the typical koloni garden owner is female and in her fifties or sixties, but a new breed like Eija are realizing the benefits of healthy living and rediscovering the beauty of these pretty little wooden cottages surrounded by space in which to grow your own food.

Eija's stuga is 35sq. m (377sq. ft) in area, decorated in stylish Swedish colours with white windows and doors and a charming balcony. Outdoor living is what this plot is all about, and garden furniture, accessories and barbecues are important. The owners who still use their plots like to eat together and entertain. Koloni gardens are used to grow fruit, vegetables, herbs and flowers in a randomly decorative rather than highly productive way, for self-sufficiency isn't a priority for this generation of gardeners.

Eija is a pastry chef who loves to grow the ingredients for her mouth-watering tarts and cakes on her allotment. Her neighbour's hen looks on.

Attached to each allotment is a pretty stuga or hut where tools can be kept safely, owners can cook their produce and even stay overnight.

Eija grows cordoned pears against a trellis, strawberries in raised beds surrounded with willow panels, and tomatoes in pots. She is a professional pastry cook and cultivates numerous herbs in containers: rosemary, thyme and sage to decorate her dishes. Her elderly neighbour grows potatoes in buckets, peas, beans and onions in square wooden beds and keeps very free-range pullets. As in most countries, newcomers – in this case Turkish gardeners – have increased the range of vegetables grown

The koloni gardens are run by associations, each with very different sets of rules and regulations: at Eija's the compost bays are only open on certain days, while many others have installed sinks and composting toilets.

Because of the Swedish climate the growing season here is short, especially in the north, and sowing outdoors is not recommended until late May because plots can be under snow till then. Greenhouses are very popular in order to get crops started under cover. Soil conditions are generally good, though, and plants grow quickly and sturdily.

HOW TO:
BUILD A PLANT LADDER

Displaying plants in small places is an art. This ladder won't take up too much space and will offer a home for your collection of herbs or even rows of salad leaves. You can use leftover or recycled timber, and the ladder takes only 30 minutes to build.

Buy your window boxes first and they will give you the width and depth of your shelves. If you'd like to make the ladder exactly as it is illustrated here, you'll need about 4m (13ft) of 150 x 25mm (6 x 1in) pressure-treated timber for the sides and shelves, plus 1m (3¼ft) of 50 x 25mm (2 x 1in) batten for the shelf supports. If you intend to paint the ladder use untreated timber – you will need to prime and paint the pieces completely before assembly. You'll also need some 40mm (1½in) screws.

1. Saw two 1.2m ladder sides with a 25° angle at the top and bottom using a set square.

2. Mark the position of the eight shelf supports equally along the two prepared ladder sides, marking the position of the final two supports at the very top of the ladder.

3. Cut the batten shelf supports to a length that will fit flush with the sides. Screw them in place using 40mm (1½in) screws.

4. Cut four shelves 25mm (1in) longer than your window boxes and screw them in, fixing the final shelf at the top of the ladder.

5. Prop your steps against a sunny wall and put your window boxes in situ. Fill with compost and plant your herbs or salads.

OUT OF THE ORDINARY

We all want food that's local and good for our health, and nothing is more local than our own gardens – but if space is limited we also can dine out at hotels and cafés that grow their own, and teach our children where food comes from. It has now been proved without doubt that eating food from the plant kingdom will improve our health, especially herbs, vegetables and fruit packed full of vitamins and minerals.

You may not have space to grow all you need, but with a little detective work you may find local initiatives such as community gardens or orchards where in return for a few days' volunteering you gain a share in the harvest or celebratory meals. Apart from the tangible rewards, working with others in the fresh air is balm for the soul as well. Any gardener will tell you that gardening lifts depression, trims the waistline and gives you pride in the food you grow.

DIGGING DOWN UNDER

Horticulturalist Michelle Shanahan cultivates the largest kitchen garden in Australia, owned by award-winning restaurant the Royal Mail Hotel and producing more than 80 per cent of their food. Nearby citrus orchards and olive groves provide fruit and oil; 120 free-range hens lay 100 eggs a day and the homestead farm raises beef and lamb. There are plans afoot to keep bees and grow mushrooms, too.

Michelle and chef Robin Wickens are constantly in touch via a computer programme that audits the produce, showing what's growing where and when it will be ready to pick. In just 30 minutes, vegetables, herbs, fruit and flowers can be harvested and on the plate.

Although the climate is on her side, Michelle has to defend her crops in order to produce food on a day-to-day basis. Using biodefensive principles, she has set up cloches to protect brassicas, spreads diatomaceous earth to discourage aphids and uses igloos for frost protection – but her first line of defence is Rocket, an Indian Runner drake and his harem of wives.

Few would claim ducks to be their first choice as co-gardeners but where there is plenty of space they make charming companions, and some breeds lay as many eggs as hens. The two main problems on the downside are the trample factor – ducks have big feet – and the mess, especially round their pond. Michelle protects vulnerable crops with fences and cloches, because anything green and sappy will be nibbled or trampled.

On the upside, she tells me they eat many garden pests, and as snail and slug predators they reign supreme; their droppings will fertilize the garden soil and their old bedding makes excellent compost when added to the heap in layers with other garden waste. They get on well with other poultry, except geese that bully them.

Ducks have the best of all worlds; they can walk, swim and fly. If you want to keep them, check with the authorities as to whether you need permission. The ideal duck home is a shed in a netted area with a pond. Keep them in a covered pen for the first two weeks until they know their address. Runner ducks won't fly away, but other small breeds may need the flight feathers of one wing trimmed; ask your breeder to do this.

Chef Robin Wickens likes his ingredients farm fresh, and a computer programme tells him exactly when Michelle's crops are ready to pick.

Using biodefensive strategies, this hotel grows the freshest flowers, the crunchiest carrots and perfect salads to feed their discerning guests.

While ducks are very hardy, house them for night-time protection and easy egg collection. They sleep on the floor, not on roosting poles; a 2.4 x 1.2m (8 x 4ft) shed with above-head ventilation would be fine for 6–8 ducks. Keep the floor dry with newspaper under straw. If you're building a pond, use a strong butyl liner and fold green plastic mesh around the turf perimeter to stop the ducks dabbling the edges away. Make sure your flock has shelter and shade from the sun.

Some breeds have been developed for their egg-laying potential – Khaki Campbells will lay an egg a day, while Pekins and Indian Runners lay 200 eggs a year and make good foragers. Heavy breeds such as Aylesbury and Rouen are destined for the table. If you have a tiny garden, a pair of small call ducks would be ideal. Duck eggshells are porous, so they have a shorter shelf life than hens' eggs. Collect them immediately, then clean and dry them.

Ducks are omnivores, and with the run of a large garden like Michelle's with access to grass, they will only need a morning and evening feed of mixed corn and duck pellets. Feed them on a hard surface or in sturdy feeders. Water is absolutely essential, and in the absence of a pond, this should be deep enough for them to immerse their heads. They'll enjoy orchard windfalls, especially after the fruit has started to ferment.

Make the most of your site, climate and soil conditions if you want to grow the most successful crops.

PETAL POWER

A dozen or so years ago, Jan and Stuart Billington uprooted their family from London and moved to a 2ha (5 acre) smallholding in Devon. Like many urban parents with pre-school children, they were looking for a better life and space to grow their own food, and Stuart's job as a philatelist gave them the financial cushion to start again in the country.

It wasn't easy, as is often the case when a complete life change is embarked upon. First they had to find the right plot with the right soil. A first option with acid soil was turned down – Jan didn't want to grow blueberries. Eventually they found a plot with clay soil that needed improvement, but would grow a wide range of plants. They converted a barn to rent out as a holiday let, planted 0.8ha (2 acres) of quince, medlar and plums to sell to local restaurants, and kept a flock of geese to keep the grass down. The geese, producers of premium, home-reared meat, supplied the top-grade fertilizer needed for their soil. Unfortunately, Jan couldn't find a local slaughterman who would come to the farm, so she reluctantly replaced her flock of honking burglar alarms with five sheep. The couple also raise a couple of pigs in their woodland and keep a few hens.

It takes a while to get to know your plot. Jan admits she spent much of the first few years in tears, at the mercy of the weather and local wildlife. Balancing nature by encouraging the right kind of predators that will keep pests at bay isn't an instant process. Nevertheless, the pair started growing vegetables and floral salads successfully and, finding that it was the flowers that sold best, they began to specialize, selling to local shops and cafés and online. Maddocks Farm is the only edible flower specialist with an organic licence, and in 2013 Jan was made a Soil Association Food Hero.

make rose water. Now their edible flowers have decorated the Beard of the Year on the cover of *Time Out* magazine, floated in the cocktails at the latest Bond movie première – an event for which Jan had to supply 3,500 flowers – and graced the presenters' tipples at the RHS Chelsea Flower Show.

Going organic had always been an important ambition and Jan has found the Soil Association to be hugely supportive, with a frontline team who are helpful and knowledgeable. She feels it's important that the public can rest assured that the flowers their food is decorated with have not been dowsed with a cocktail of agri-chemicals.

We all love flowers for their colour, shape and fragrance, but it's only recently that we've been eating them as well. 'Flowers should be appreciated as an ingredient in their own right,' says Jan. She advises that for a citrusy flavour, add oxalis, tageles and begonias; mustards, wild rocket and nasturtiums add a spicy bite; chives and wild garlic taste oniony; tulips and gladioli give a sweet, crunchy texture; dahlia and snapdragon petals taste of chicory; carnations and dianthus add a touch of clove and calendulas a teaspoon of nutmeg. All the herb flowers taste like their leaves and all are beautiful additions to a cook's ingredients.

To give them a good start, their seeds are sown in modules inside one of their nine polytunnels, then planted out in rows in no-dig raised beds mulched with their own compost, interplanted with insect-attracting calendulas and nasturtiums. More than 200kg (440lb) of rose petals are grown to

Growing under plastic is an effective way of keeping pests and diseases at bay, and beats bad weather.

HOW TO: MAKE A GARLIC SPRAY

I grow lilies in pots. In the past, they have been decimated by lily beetle, a rather beautiful bright red insect whose larvae completely strip the leaves and buds. Last year I sprayed my lilies with garlic steeped in water and they thrived. Luckily, the smell of garlic was overpowered by the scent of the lilies!

I sprayed my lilies daily every evening, avoiding the flowers and taking care not to spray near my eyes or face, and made sure the bottle was kept out of reach of children.

Other gardeners believe that garlic water combats aphids, and some add a few drops of chilli oil or a squeeze of liquid soap. Chilli oil must be used with care, as it can cause skin irritation. Never allow it near your eyes.

1. Crush three garlic cloves, using a mortar and pestle. Put the crushed garlic in 568ml (1 pint) of water.

2. Shake the bottle and leave to steep for two days.

3. Thoroughly clean a pump-action spray bottle.

4. Strain the garlic water into the spray bottle, attach the spray fitting tightly and then store the garlic water in the fridge.

HERBS FOR HEALTH

Mockingbird Meadows is a haven for endangered medicinal plants, a refuge for the local bee population and a welcoming oasis where people can learn about the healing properties of herbs. A small family living off just 1.4ha (3½ acres), designated as a United Plant Savers Botanical Sanctuary, this homesteading farm in Central Ohio produces honey and herbs, and through workshops and visits gives everyone the opportunity to connect better with food and feed their souls too.

The owners are Carson and Dawn Combs, the former a biodynamic farmer and the latter a community herbalist and ethnobotanist. When they first started beekeeping a pair of mockingbirds built a nest in the trees above their first hive, watching over their endeavours and, legend has it, protecting the bees. As this is a Certified Bee Friendly Farm, the bees' health comes first, and their colonies are never fed on sugar water or corn syrup – just on honey, which is their right.

Special herbal blends are planted for their bees, sited near their hives, and they're disturbed as little as possible. Their honey is harvested just once a year and their apiary community is built from natural swarms rather than by importing outsiders.

Mockingbird Meadows' colonies of bees have special herbal blends planted near their hives.

The Combs believe their raw honey has many allergy benefits, and use it combined with herbs to make salves that help with problems such as insomnia (blended with hops, passion flower and skullcap); to induce calm and relaxation (mixed with nepeta, lemon balm and Californian poppy); and to settle the stomach (combined with wild yam, ginger, spirulina, peppermint and lemon balm).

We all know the glorious flavours that culinary herbs provide to enhance our cooking. Where would a kitchen be without the basic Mediterranean bouquet garni of rosemary, sage, thyme and bay? Herbs are the ultimate Swiss army knives of plants in their multiple uses: beautiful to look at, hugely beneficial to pollinating insects, a multi-use pharmacopeia and indispensable in cooking, gardening, cosmetic and historical contexts.

But Dawn is most interested in what herbs from our gardens, smallholdings and hedgerows can do to help us keep healthy. Just steps from her own front door, she can pick hops for their sedative, relaxant and anti-microbial properties; hawthorn to help with circulatory problems; lemon balm to calm restlessness and help sleeplessness; and golden rod as an anti-inflammatory. The abundance of hedgerow food available for foraging never ceases to amaze her: flowers and berries, nuts and seeds, stems and roots, fruit, leaves – all evidence of nature's bounty.

As Homesteaders of the Year in 2013, and regular award-winners for their honey at the Slow Food Festival, the Combs are clearly getting things right. Dawn says, 'I love walking out on to my land and finding everything I need for my family.' With a pond, three barns and a greenhouse, they keep very free-range poultry for eggs and dairy animals for milk, cheese and butter: the smallholding is their larder and a playground for their two small sons.

Dawn's books on herbal healing and her blogs, videos and newsletters on her website (heallocal.com) are packed with useful advice; her ethos of using local medicinal herbs to heal and the benefits of terroir for fresher food, plus encouraging a greater awareness of plants in our own backyard, is persuasive and thought-provoking.

Hippocrates said, 'Let food be your medicine, and medicine be your food.' Dawn and Carson believe wholeheartedly that good soil is the source of life, resulting in healthier plants that in turn provide better health to the animals and insects that eat them; and that we are the ultimate benefactors of this thoughtful farming ecosystem.

Join together with your friends and neighbours to grow and pick fruit and celebrate the year's bounteous harvest.

ORCHARDS FOR ALL

If you only have room for one tree in your garden, plant a fruit tree. You'll have blossom in springtime for the bees and fruit in the autumn to share with the birds. Even the smallest space could house a dwarf pear tree in a pot, an espalier cherry against a wall, a cordon apple instead of a fence or even a boundary fruit hedge packed with damsons, plums and crab apples.

Fruit was once grown commercially all over the UK, and every farm and large garden had its own fruit trees surrounded by edible hedgerows. Farmhands were paid in home-brewed cider, and local children scrumped healthy treats. Fruit is easier to grow than vegetables, nicer to eat and prettier to look at – but if you don't have space, why not join a community orchard near you?

No matter where you live, there are alternative ways to grow and harvest. The Urban Orchard Project and Common Ground work in partnership with communities to plant and restore orchards in towns and cities and help us all to rediscover the pleasure of eating home-grown fruit, creating, as they put it, 'lush cities across the United Kingdom swathed in fruit and nut trees'.

Children from the local primary school learn where their food comes from, and what goes into producing it.

Both bodies need more volunteers – the more people involved, the easier it is to look after the trees, and during their first three years all fruit trees need a bit of attention. Trees are best planted bare-rooted when dormant, so it'll probably be a chilly day but a bit of digging warms everyone up.

To plant a fruit tree successfully:
- Mark out a square of about 1m (3¼ft) with your spade.
- Dig down about 5cm (2in) to take off the turf.
- Dig the pit a spade deep then break up the base with a fork. The pit needs to be wide enough to accommodate the root ball comfortably.
- Ask someone to hold the tree while you backfill with soil you've dug out.
- Make sure the graft is about 10cm (4in) above the soil level and tamp down the soil.
- Mulch with a 10cm (4in) layer of compost, keeping it away from the trunk.
- Add a tree guard and mulch mat to stop weeds encroaching for the first three years.

Over the last half-century more than 60 per cent of traditional UK orchards have disappeared, losing vital habitats for lichens, mosses, insects and birds, as well as ancient fruit varieties. Many farmers have grubbed up their orchards, replacing standard trees with dwarf and more easily grown varieties, but some have been sold and snapped up by community groups.

In 1972, a 22.3ha (55 acre) traditional apple orchard, Standen Fruit Farm near Benenden in the High Weald in Kent, was sold off in 320 small parcels of land. Luckily, planning regulations prevented development, but the trees were neglected and the site became derelict, a blight on a beautiful area. In 1990 a group of local people got together to restore and replant, prune and learn together, with help from Natural England and the Soil Association.

Any situation where a community grows,
harvests and gets together is a celebration,
and a chance to raise funds.

It has taken time to sort out teething problems, but over the last few years a charitable trust has turned the orchard into an asset for the whole village. The grazing is let out so flocks of Lleyn and Suffolk sheep keep the grass down, contract workers help to harvest the fruit from the thousand or so trees, and the committee of villagers, including a journalist, software designer, builder, farmer and mature student and their friends and families, is involving more and more local people in the joys of working outside.

As committee member Kent Barker told me, 'The orchard holds open days with music and barbecues advertised in the village newsletter, inviting people to pick, eat and juice apples in the communal juicer – some for juice and others for our much appreciated cider. We ask villagers to bring their own bottles and step ladders.'

Pruning ancient standard fruit trees is a specialist art requiring chainsaws and planning, but there are basic rules:
- Prune pip fruits in winter and stone fruits in summer.
- Make sure you have sharp, clean tools.
- Step back occasionally and assess what you've been doing.
- Never remove more than a quarter of a tree's wood in a year.
- Don't leave stubs as they'll rot. Cut back to the base or to a side shoot.
- Prune out all dead, damaged or diseased wood.
- Prune to encourage ventilation and produce more fruit.

Turn your fruit into refreshing juice and your juice into delicious cider.

At last, grants have financed new trees, secateurs, chainsaws and a tractor; the footpaths are being maintained so the public can visit and admire, and the school has adopted trees and schedules visits to view their progress. A section of woodland is being preserved as a wildlife sanctuary with bramble patches left for nesting birds to protect them from buzzards that circle overhead.

A monthly calendar of activities to keep the orchard in good heart has been set out, so newcomers can learn on the job – the best way. In a community orchard no one expects you to have had previous experience, and all orchards have their own style. If you feel the work involved may be a little too physical to start with you can pace yourself, since all sorts of skills are needed.

HOW TO:
MAKE FRUIT LEATHERS

Fruit leathers are a lovely treat for children – they're very moreish, and easy to make. Use fruits that are high in pectin: apples, plums, currants and gooseberries do well. You can add oats or ground almonds for more substance and sweeten them with sugar, syrup or honey. Add cinnamon or vanilla for extra flavour.

Once dried in the oven, the leather will be ready when you can peel a corner easily from the foil. Cut into strips or shapes with a cutter and store rolled in baking parchment in an airtight jar in a cool dark place for up to six weeks, but check for mould before eating. Mostly, you'll find they are eaten up instantly.

1. Top and tail 0.8kg (1¾lb) blackcurrants and put them in a saucepan. Add 100ml (3½fl oz) water.

2. Simmer on a low heat, stirring occasionally. Sweeten with sugar to taste.

3. Mash then sieve the fruit, removing all the pulp.

4. Pour into a tray lined with foil or ovenproof clingfim and dry in an open oven at 60°C/140°F/Gas ¼ for 4–6 hours.

FIRST STEPS TO KEEPING ANIMALS

Hen Corner is a little bit of country life in the city. Since moving to their tiny terraced house in outer London, Sara Ward and her husband Andy have recreated a tiny rural utopia with an orchard, chicken runs, veg plot and beehives, and they spread the word with evangelistic zeal, holding courses, getting involved with community projects and living their country idyll to the full.

Whether it's keeping poultry, making cider, growing your own food or learning artisanal cooking skills, Sara will show you how. Group classes are small and intimate. They take place in her house and garden and often involve a good deal of sampling delicious home-produced food and drink – great incentives to tempt you to prepare them yourself.

Lessons learned in childhood take root and grow, and as vice-chair of governors of her local primary school, Sara has years of experience working with kids in school and in the voluntary sector. She loves going into schools and taking part in projects involving life cycles, keeping hens and bees and food production. Teachers are pleased to see the countryside coming to school and feel that Sara is bringing the curriculum to life.

Children who visit her productive garden to learn where eggs and honey come from enjoy holding and patting docile chickens or dressing up in protective clothing to collect honey from the hives, and may even turn into the smallholders of the future. Learning in a real-life set-up while you're having fun is the best way to get the message across.

These children will never forget where their food comes from and develop ethical shopping and eating habits in the nicest possible way: from the heart.

Anyone wanting to learn how to keep animals should take the time to find out what's involved – go on a course with a specialist, spend time on a farm or smallholding and visit breeders. At the very least, buy a good book on the subject, watch online videos or join a breed club to meet others who keep stock successfully. Immerse yourself in knowledge before you make the decision and buy.

Ask yourself first: is your garden or smallholding large enough to sustain stock? Will you have the time to look after them and give them a good life? As owners we have the responsibility to look after our charges properly and they have the legal right to the following welfare needs:

Having fun together is the quickest and easiest way to learn.

- A suitable environment – animals need enough space to live happily in the right type of housing. Some may need special housing that might need planning permission, or a particular type of fencing.
- A suitable diet – find out about feeding and drinking requirements and where to buy animal feed and equipment.
- To be able to exhibit normal behaviour patterns – learn how your animals live day-to-day and help them have a good life.
- To be housed with or apart from other animals – most farm animals need to live in groups, which means at least one other of the same species that get on well together.
- To be protected from pain, suffering, injury and disease. Make sure you

find a good vet nearby. Animal welfare groups strongly recommend you don't slaughter animals yourself without training as you could cause suffering, and that would be an offence.

If you have only a small garden and want to keep chickens, buy a couple of bantams and try to let them range free for a few hours a day; if you don't have much water on your land but want to keep ducks, call ducks are the answer. Pygmy goats and pigs could be a first step towards their larger relatives, but whatever size animals or birds you decide on, their welfare should come first, and everyone will live a happier life if you know how to look after them well.

HARVEST TIME

The passing of seasons is significant in the smallholding year, and if you plant well and the weather is fair, in due course you'll harvest bounteous crops. It's a busy time – the culmination of all your efforts.

For Liljan and Roland Bengtsson-Johansson in Sweden, it's crucial to harvest top-quality new potatoes in time for Midsummer Day parties, while for Will Davenport at Davenport Vineyards the results won't be obvious until the vintage is tasted. But whatever you plant, harvest is a time of plenty and an occasion to celebrate together.

My family like to eat our produce out of doors, and we use our plot as an outdoor dining room; community gardeners can enjoy learning how to make jam together; and Frances Avery works hard to preserve all the food her family smallholding produces for the winter. Harvest is the best and busiest time of the smallholding year. Enjoy!

Every spring, as the soil warms up, these Swedish smallholders plant their precious potato crops.

SCANDI SMALLHOLDING

Sometimes the lie of the land, the type of soil and the climate dictate the ideal crops to grow. Liljan and Roland Bengtsson-Johansson were lucky to inherit their family farm on a south-facing slope along an inlet of the Kattegat near Varberg, south-west Sweden, in an area famous for its potatoes. Now in their late seventies, they specialize in growing superb new potatoes to harvest for Midsummer Day, when, cooked with dill, they grace the party table – a national speciality.

Most of their 12ha (30 acres) of land is rented to a neighbour who keeps cattle, but generations ago, the family realized that the light, sandy soil on their top fields was perfect for potatoes. Every spring, as the snow melts, Liljan and Roland chit and plant the earliest seed varieties, Rocket and Swift, and cover them with fleece. As their potatoes grow they earth them up to be ready in time for the midsummer celebrations.

Later varieties are put in as the soil warms up. Roland makes the furrows with his beautifully restored vintage tractor, while Liljan carefully plants the potatoes, scattering in a little fertilizer before covering them with soil – a labour of love for a crop they know like the back of their hands. Every year, 600kg (1,320lb) of new seed potatoes are brought in to grow healthy plants, but the couple don't rotate their crop. Maybe the freezing winters kill off any diseases and pests, but in most climates, it would be essential to rest the potato fields or to spray them with fungicides and pesticides.

The Bengtsson-Johanssons sell their famous crop from the farm gate, along with home-laid eggs and an impressive array of vegetables: carrots, beetroot, onions, various types of beans and courgettes, started in their greenhouse. Their small orchard supplies plums, apples and pears.

Their land also includes a small forest, which supplies the farm with logs, cut to size especially for their boiler and neatly stacked – a common sight in the Swedish countryside. With their small flock of horned sheep and milking goats that occupy a beautifully restored barn, fed on home-cut meadow hay, Liljan and Roland live a comfortable life in a style unchanged for generations.

Selling from the farm gate is a good way to market crops. Customers can see where their food comes from, while producers get valuable feedback.

Liljan is very fond of her goats and believes they are ideal animals for smallholders, providing milk, meat, manure and wool. Goats like cold, dry conditions, spending Swedish winters inside their barn, and need shade in summer and shelter from rain. They are herd animals, so shouldn't be kept alone, and browse vegetation and hay, rather than grazing grass; poisonous plants must be removed from their pasture. They get on well with sheep, though they are considered to be much cleverer and more adventurous, so durable fencing is a must.

For milk goats, choose a breed such as the Saanen, Nubian or British Toggenburg, which have good yield, long lactation and high butterfat content, so you can make cheese and yoghurt as well. To keep the milk supply flowing, you need to breed your nannies once a year in the autumn from eight months of age. Stop milking once the goat is pregnant, but resume when the kid is a week old. Excess billy goats can be raised for meat. But before you acquire your goats, go on a course to learn how to breed and raise livestock.

Feed your flock on hay and concentrated ration mix, increasing the mix when they are pregnant or lactating. Goats need regular hoof care, with hoof trimming every 6–8 weeks, as well as parasite control. Keep a medicine book to log problems, and remember that in the UK, holding numbers must be obtained and your goat should be tagged.

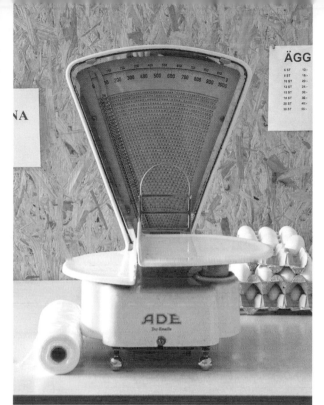

My grandson Ludo counts his apples, picked from trees and fenced cordons.

FAMILY ENTERTAINING

When I moved from my 0.4ha (1 acre) Suffolk garden to the seaside in Kent, I wanted a garden to share with my family: two sons Jacques and Max, their wives Saskia and Helen, and grandsons Ludo and Etienne. So I decided to divide my 45.7 x 15.2m (150 x 50ft) plot into three: the end part is a wild garden, the middle is a small orchard, and the third nearest the house is an outdoor kitchen.

I planted my orchard with cherries, apples and pears, apricots and greengages, quince, damsons and persimmons because Kent is famously the Garden of England and this whole area used to be an orchard that ran down to the sea. Also, fruit is fun to grow, delicious to eat and wildlife-friendly. An orchard is also an ideal place for hens to scratch about.

Orpington bantams are a tiny, friendly breed and my trio live in a henhouse made by Jacques in a large fox-proof weldmesh run. They're only allowed into the orchard when I'm with them, because bold urban foxes visit. Descended from Asian jungle fowl, domestic chickens need a shady area. They like to potter in long grass, eating seeds and insects, but it's short grass that's protein-rich, so I've mown paths through the orchard. My bantams finish off fallen fruit, especially the pips, and even eat windblown blossom petals.

They're fed a scoopful of layers pellets every morning and one of mixed corn at bedtime, with access to clean water within their run. Their nestboxes are cosy with dried moss from the lawn, their house is lined with old newspaper for easy cleaning and their purpose-built dust bath under the house is full of wood-ash from my woodburning stove. They have a good life!

All my family has autumn birthdays and most are celebrated out in the garden.

People keep poultry for various reasons. Some prefer hybrids for their mammoth egg-laying abilities; others, such as Sheila Hume (see pp.112–115) have chosen particular breeds for the colour of their eggs, and yet others want to give battery hens a new life. Mine are garden hens, kept for fun, for their charms and for their good behaviour in the garden. Birds that lay a lot eat a lot too, and if what's on offer is your garden, soon there'll be barely a blade of grass left. Breeds that make good co-gardeners include Brahmas, Silkies, Pekins, Belgian Bantams and Orpingtons. You'll also be preserving these pure breeds for posterity and for their particular traits that have been used to develop modern hybrids.

Divided from the orchard by a cordoned apple fence is a tiny veg and herb garden where I grow seasonal delights. I use a wooden Kilner juicer to make fruit juice to freeze, we make jams and jellies, and fruit purées fill the freezer. My grandsons and I pick rhubarb from April, then gooseberries, wild strawberries (white ones that the birds seem to leave for us) then currants – especially jostaberries (a gooseberry/blackcurrant cross). We stuff ourselves with cherries, damsons, gages and apricots, followed by apples, pears and passion fruit, until the season ends in November with persimmons.

Our outside kitchen houses two permanent tables, various benches and several folding chairs. Sturdy outdoor garden benches can instantly be made comfortable with cushions, and tables covered with pretty cloths. Add a bunch of home-grown flowers and herbs, and you have the most hospitable venue for family celebrations. We've welcomed new arrivals in darkest November with bonfires, hurricane lamps and sparklers, celebrated birthdays, anniversaries and holidays, and, sharing familiar menus, remembered those sadly missed.

Food cooked out in the open, either in an outdoor oven, a fire bowl or garden fire pit has a special flavour.

A small terracotta pizza oven for slow-cooking sits high on a long work surface, where we can bake all day once it has been fired up, while more immediate grilling is done on a large metal fire bowl. We also use this fire bowl to keep us warm on chillier evenings. In my previous garden, I had an open-air fire pit that doubled up as a bonfire site and the chickens' favourite communal dust-bathing venue, and I shall build another one here.

Whether it's cooked on a campfire with a tripod or on an outdoor wood-burner, such as a rocket stove (where you can cook a meal with just a few twigs), everything tastes better cooked outside. There's an endless supply of kindling, prunings and small wood in most gardens – though it's worth keeping useful bits somewhere dry – and I love the different flavours you get from cooking on wood. Try grilling salmon on alder wood; apple and cherry are mild and good with poultry; burn maple with ham and oak with beef (but don't cook with pine or any wood that has been treated, including tanalized timber). You may want to progress to an outdoor charcoal hot smoker which will open up a new range of flavours.

With easy access to the indoor kitchen, I often take the simplest meal outside to eat – an omelette from my hens' eggs and a salad from the veg plot – and enjoy picking, cooking and eating meals grown just a few feet from the table. This is an easy, sociable garden: friendly to local wildlife, ideal for my hens, productive and fun for the whole family.

My grandsons and I pick fruit from springtime rhubarb to the last of the apples. Foraging along local lanes for blackberries is one of most people's favourite childhood memories.

HOW TO: MAKE A FIRE PIT

A fire pit is the easiest way to cook food in the garden or on the allotment. It's also a nice way to keep warm on a chilly evening. Pick a site that's away from overhanging trees or any other fire risk, but somewhere that's good to hang out and relax.

Remove the turf from a circle with a spade – leave it face down in the compost heap, where it will make good soil – and then dig or pick-axe down to the depth of two or three courses of old bricks. Stack them round the inner edge, overlapping the gaps in between the bricks with the next course. It isn't necessary to cement them in, but you can if you want to make a permanent site.

Make a fire in the pit, using garden kindling, seed heads or pine cones, then add charcoal or smoking wood chips if you want to add flavour. I always save any herb twigs or branches for a special treat. Cover with a metal grill. Use your fire pit to slow-cook in heavy pots or to quickly grill kebabs with vegetables straight from the garden.

1. Pick a safe site to dig your pit and remove the turf.

2. Dig down and line the edge of the pit with second-hand bricks.

3. Cover with a metal grill.

4. Make a fire using garden twigs and branches. Add charcoal to grill or to slow-cook produce in a pot.

*Find a welcome as a garden volunteer,
buy local produce and learn a skill.*

GROWING TOGETHER

Not everyone has the luxury of garden space, especially town- and city-dwellers. You might be lucky enough to rent an allotment, but waiting lists are long and not everyone can commit to the hard physical work and spare time needed to work a whole plot. Newcomers to gardening often lack the confidence to start in full view of others, too.

An alternative for those who like to dip into gardening from time to time and learn as they go is a local community garden. Go online and check websites to find one near you, or see if a local school, church or old people's home needs help with gardening work. Maybe someone in your neighbourhood would like to share their garden in return for some of the produce that you grow; some Transition Town websites put people in touch with those with land to share. You'll relish the harvest, and enjoy gardening alongside others, learning as you go.

My local community garden used to be a disused allotment area in the heart of town and over the past five years, members have slowly turned the plot into a lovely garden. Anyone can join in. Whether they're knowledgeable gardeners, want to get fit or just sit and watch, all are welcome. The garden also supports those who want to develop basic skills in horticulture, offers them a reference and has recently opened a bodgers' hut where joinery, turning and carpentry are taught. With a pond, children's area, wildlife garden and lots of raised vegetable plots, there's something for everyone.

Food preserving techniques like jam- and jelly-making, bottling and pickling will disappear for future generations unless we preserve their recipes too.

On the day of our visit, a jam-making session was in full flow in the open-air classroom and community gardener Becky Richards was teaching visitors how to make the most of surplus produce. Strawberries, raspberries and currants were being turned into jams, jellies and coulis. Some crops had been picked from the Stream Walk fruit cage or from the boundary hedge where wild fruit grows, while others had been grown at home by gardeners who brought them in, wanting to know how to preserve them. The results were to be sold to raise money for the garden. I shared my recipe for fridge jam: using up small amounts of fruit, adding sugar to sweeten, not preserve, and keeping it fresh in the fridge for a week or so.

What to do with surplus produce is a problem for all gardeners, solved sometimes by growing fewer crops in succession, but the vagaries of the weather and growing conditions often make it impossible to plan ahead. There are several different ways to deal with gluts:

- Pickling in vinegar, including relishes, chutneys and ketchups.
- Preserving in sugar to make jams, jellies, sauces, cordials and sweets.
- Curing in salt and smoking.
- Air drying, dehydrating and fruit leather-making.
- Preserving under oil, fat and butter.
- Preserving in alcohol.
- Fermenting.
- Heating and bottling.
- Freezing.

Other activities on offer at Stream Walk include creating a forest garden, making paths, laying hedges and making compost. Some community gardens keep bees, hens and even pigs, with maintenance and produce divided among their carers. On Friday mornings at Stream Walk a group of mums and toddlers come to learn and play, and Saturdays are open days for anyone to participate or just enjoy a cup of tea from the new tea hut.

As an occasional volunteer, I receive emails telling me what jobs need doing, from planting native trees destined for local open spaces to pond clearing, or inviting me to a barbecue and keeping me up-to-date on plans for the new compost loo. The whole garden is also available as a venue for local celebrations, further spreading the word about the joy of home-grown food.

A TOAST TO ORGANIC WINE

It takes 600–800 grapes to make a bottle of wine. Will Davenport planted his first vineyard in 1991 – a tiny 2ha (5 acre) plot that, four years later, produced 900 bottles of wine. That's not a bad harvest, considering he was working in another vineyard 160km (100 miles) away at the time. Since then, what was a weekend hobby has grown into one of the country's top organic wineries employing six helpers and in 2015 producing 30,000 bottles from several sites in the south of England.

Trained in Australia with vineyard experience in France and California, Will admits that while growing grapes for wine can be stimulating, growing them organically is a challenge. The grape isn't native to Britain, so aspect, location and soil type is of paramount importance. If you have a sunny, south-westerly facing slope that's well drained, and less than 91.4m (300ft) above sea level, with a little wind to blow away the moulds and mildews, you're in with a fighting chance.

It can be done, you can produce good wine, but not every year, thanks to the vagaries of the weather. Will's blogs tell of good years and bad: '2014 will go down as a really good year: warm spring, no late frosts, dry during flowering, hot July, wet in August (OK, a little too wet) and one of the driest Septembers on record. It got colder by mid-October, but yield was high across our grape varieties and sugar levels were acceptable.' And then: '2013 has been a year of difficulties: a cold late spring, followed by a hot dry summer and the wettest autumn in five years.'

The grape-picking season gets underway at Will Davenport's vineyard – with top vintage results hopefully.

WINE	VOL	TK	RACK	YEAST	SUGAR	SWEET	PN MT	SO2	ACID	
FO	1180	39	✓	✓	✓	X	X	O		O
LS	2120	36	✓	X	X	X	X	O		O
FB	585	14	✓	✓	✓	X	X	O	A	O
LO	1550	1	✓	X	✓	X	X	O		O
LDF	1135	01-6	✓	X	X	X	X	30		O
LB1	3570	37	✓	X	✓	X	XW	O		O
LO	1710	F1	✓	X	✓	✓	X	O		O
FP4	1195	29	✓	✓	✓	X	X	O		O
LB2	890	2	✓	X	X	X	X	O		O
FC	800	8	✓	✓	✓	X		O		O
LH	1020	3	✓	X	X			O		O
LB2	750	6								

The grape isn't a British native, so not every summer produces good wine.

You'll need professional advice to start (there are regional vineyard associations ready to help), but a hefty bank balance is needed to install trellising and plant 4,000 vines per hectare (2½ acres) and it will probably take ten years to turn a profit, especially if you're investing in machinery for your own winery.

Like most smallholders, Will has diversified. He keeps a small flock of easy Wiltshire Horn sheep, an ancient breed with great resilience that has only survived thanks to the protection of the Rare Breeds Survival Trust. Nowadays wool production has become uneconomical and there are costs involved in shearing, dagging and dipping, so it's a big advantage that the Wiltshire Horn sheds its fleece naturally. The large, goat-like ewes are also suited to outdoor lambing.

The sheep roam 10ha (25 acres) of small fields following centuries-old boundaries. Their grazing is managed to enhance biodiversity and the farm is registered under the Environmental Stewardship scheme. Wild flowers are encouraged, woodland is coppiced and Will is digging a shallow scrape pond. Kestrels and grass snakes abound. The site is home to a collection of beehives, with their occupants helping to pollinate the grapes and providing honey that is sold on-site. Will also grows vegetables for his family of six.

Will grows vegetables for his family of six and keeps bees for honey as well as growing organic grape vines in his vineyard.

The vineyard uses home-made compost and green manures to increase fertility. The grape skins and stalks and winery waste are added to green council waste and composted. No chemical pesticides are used, only organically approved fungicides – currently down to half the permitted sulphur dioxide. Only clay fining agents are used in the winemaking process.

Pinot Noir, Auxerrois, Bacchus, Faberrebe, Chardonnay and Pinot Meunier are grown organically to produce an award-winnng wines, such as the sparkling Blanc de Blancs, described as 'a dry style with fine bubbles, lovely peach and white fruit flavours on the tongue – a fantastic alternative to Champagne'. Forty per cent of English wine sold is sparkling, but Davenport Vineyards also produce a rare red Pinot Noir and a white described as 'kiwis and greengage on the palate, crisp but nicely balanced. A zesty, spicy dry white that's a consistently persuasive advert for English wines'.

A TIME OF PLENTY

There's a season for everything: a time to sow and a time to harvest, and even if you crop carefully in succession, harvest is a time of plenty. Frances Avery deals with the produce from a large family smallholding: an 0.2ha (½ acre) vegetable garden, a fruit cage, two orchards and a nuttery – 2.8ha (7 acres) in all, farmed by her husband William, their two sons, Oscar and Rupert, and their daughter Amélie. In high summer she handles daily boxes of fruit, nuts and vegetables.

Frances loves sowing crops, especially in pots and seed trays in her greenhouse, then cooks up a storm to make the most of what grows, while her family dig, water and cultivate. She preserves what she can for the larder over winter and then sells leftovers at the farm gate, to make pocket money to buy next season's seeds – a seasonal life providing for her family that gives her great pleasure.

The vegetable plot is divided into long 1.2m (4ft) wide beds intersected with paths wide enough to take a wheelbarrow. There's a larger netted bed for soft fruit and another used as a permanent home for artichokes and asparagus. The remaining four beds are rotated for potatoes; courgettes and squash; peas and beans; and leafy veg. Tomatoes, peppers, chillies, aubergines and cucumber are grown in the greenhouse.

Frances deals with all the crops her family grows, preserving fruit, vegetables and nuts so that they last over winter until the next year's harvest comes along. She pickles, jams, freezes and bakes large amounts of produce with great skill and limitless patience.

One of the orchards is home to cherries, apples and pears, and the other is full of plums, mirabelles and greengages. Walnuts, hazels and cobnuts are grown in the nuttery. Nuts are a good way of producing home-grown protein and can be dried and kept in nets, while many varieties of apples and pears can be stored by wrapping them separately in newspaper and placing them in cardboard or wooden trays kept in a cool, dry, rodent-free space. Most other fruit needs to eaten straight from the bush or tree, or preserved in some way.

Frances occasionally feels overwhelmed by the trugs full of produce that land on her kitchen table, but after she has frozen, puréed, made into jam or pickled as much as she needs she sells leftovers to a local farmers' market. The main ways to preserve food are with vinegar, sugar or salt; with air, oil and alcohol; or with heat or cold. Curing, smoking, bottling, fermenting and drying, to say nothing of freezing, pickling and conserving – these are all time-honoured skills that have never been more needed in this fast food, throwaway society. They take time, but can be built into a busy lifestyle, reminding you of the joy of harvest all winter through.

The Averys' beans, broccoli and leafy greens are blanched and then frozen in family-sized portions; root vegetables are left in soil clamps; squashes and pulses are dried; onions and shallots are plaited and hung; and the rest are preserved. Frances makes numerous jars of chutney and pickles. Tomatoes, rhubarb, courgettes and green beans are chutneyed with sugar, spices, dried fruit and vinegar, while gherkins, tiny onions, cauliflower and beetroot are all delicious pickled in good vinegar flavoured with added herbs. Vegetables such as artichokes, peppers and tomatoes, can be preserved covered with oil, salted in brine, or smoked, while greenhouse crops can be sun-dried or turned into harissas, pestos and ketchups.

Fruit is mostly frozen whole for tarts, pies and clafoutis, or puréed for sauces, fools and coulis, but some is made into jam, jellies or leather, soaked in alcohol or turned into cordials. This is a good way to make winter drinks which can be frozen in small plastic bottles. Apples and pears can be pressed and pasteurized into juice or fermented into cider, perry or country wines.

Plums from the orchard are her biggest crop and her clafoutis and tarts are legendary. She takes any surplus to her local farmers' market.

HOW TO: MAKE PLUM GIN

Plums, damsons and mirabelles ripen all at once, but steeped in gin in a pretty bottle they are long-lasting and make perfect Christmas presents. Any excess fruit can be preserved in this way: rhubarb with orange in vodka, grated quince in brandy and of course sloes in gin.

Select ripe fruit and wash it well. Don't remove the plum stones, as they add a nice almondy flavour to the gin. Fruit with a thick skin should also be pierced (traditionally with a silver pin), but a few days in the freezer will have the same effect. You'll need a wide-mouthed jar that has been through a dishwasher cycle or washed and dried in the oven at 140°C/275°F. When your gin is ready, label with the vintage – the leftover boozy fruit can be kept and eaten with ice cream for a treat.

1. Pick and wash 300g (10oz) of fruit, then pierce the skin with a fork or silver pin and transfer to a prepared jar.

2. Add 80g (2¾oz) of caster sugar – you can add more later if you prefer your drinks to be sweeter.

3. Top up the jar with 700ml (1¼ pints) of gin, pop on the lid and shake to dissolve the sugar.

4. Store in a cool, dark place for 3–4 months, shaking the jar and tasting for sweetness every now and again. Strain through a sieve into a bottle, using a funnel.

GARDEN FARMING

Learning to grow plants and raise animals takes time. Many schools today have farms where the knowledge gained can be translated to academic subjects that are part of the curriculum, while some children grow up in farming families. For most of us, though, farming is something we come to later in life. If you are planning to keep animals of any description, it's very important to first go on a course to find out how to give your charges a good life.

The lie of the land, the type of soil and obviously the space you have will dictate the crops you plant and the animals you keep. Pigs, for instance, need a light soil; bees must have access to flowers; and flowers need the right soil to flourish. Planning beforehand is a must, and visits to model farms and successful smallholdings can point you in the right direction.

THE REALLY GOOD LIFE

Karen and Jeff Nethercott live the good life. Ex-townies, they moved to Norfolk a dozen years ago to become traditional smallholders. As Karen says: 'It may seem unromantic, but the reasons for our move were food-based. I wanted to make sure we were eating produce from animals that had lived a good life too.'

Finding the perfect plot wasn't easy. They scoured the Internet and smallholding press, went to auctions and visited many properties. They joined their local smallholding group – a network of passionate garden farmers, generous with their knowledge. The Norfolk farming community is full of houses with a few hectares as many people have sold the majority of their land to the big boys, leaving ideal smallholdings.

Eventually the Nethercotts found a house with 2.8ha (7 acres) of land, encompassing two large fields bordered with native hedgerow – larders of bullace, hazel, sloes and kindling. They divided their fields into sections with electric and stock fencing to rotate their animals, planted an orchard of apples, quince and medlars, and laid out a veg plot of 20 raised beds near the house.

With all the versatility of the traditional smallholder, Jeff built a small barn from timber clad with corrugated iron to house their stock and feed, and a shed and garage using the same materials. Their ingenious compost heaps recycle household waste and all their animals' bedding and manure.

Originally, realizing their planned flock of sheep wouldn't keep up with the grass growth, they had their fields topped (mown) by a contractor to avoid buying costly haymaking equipment, but since then they have bought a 1954 grey diesel 'Fergie' tractor with a topper so they can cut their own hay. Jeff loves his elderly machine, cheap to buy and easy to repair, even though he is deaf for hours after using it. The couple's only other bit of kit is a livestock trailer with an internal gate.

Karen doesn't need an excuse to scour sales and farm auctions for feeders, drinkers and hurdles. Builders' buckets can act as sheep feeders, children's paddling pools suffice for ducks, while metal-lidded dustbins are ideal for storing 20kg (44lb) bags of feed out of reach from rats. Second-hand equipment can be found in local newspapers and online too.

Karen and Jeff started with chickens. A lot of people do. They knew hybrids to be more reliable layers, but prefer characterful pure breeds: big blowsy Orpingtons and Brahmas, neat little Pekins and blue egg-laying Araucanas. They then added a few Runner ducks and a pair of Emden geese to make up their flock of ace pest controllers that live in the orchard.

Karen is particularly interested in old breeds of farm animals that are under threat; the Rare Breeds Survival Trust website is a good place to view your options if you are of the same mind. Seduced by their striking appearance, Karen wanted Lincoln Longwools, but it was hard to find a serving ram locally so she settled for Ryelands. With two other flocks nearby, she sends her ewes on holiday for a few weeks in autumn, and they come back 'in lamb'.

The Rare Breeds Survival Trust advises on the best breeds of livestock for farmers, and protects them for agricultural history.

Ryelands have been around for seven centuries. Stocky, docile and easy to look after, this is an ideal breed for smallholders. They make good mums, do well on a diet of grass without the need for additional feed, rarely suffer foot problems and produce tasty meat. Karen and Jeff sell their lambs as hoggets, which means they've kept them for between 12 and 18 months. This is the firmest, tastiest choice of sheep meat and it's hard to come by; rarity makes it a premium foodstuff, and reaps rewards for the smallholder. A good-sized lamb will give you 15kg (33lb) of meat.

Producers should take every opportunity to sell their wares: on websites, at markets and at the farm gate. The public increasingly wants to know the provenance of food.

When it comes to pigs, the Nethercotts have kept Tamworths and British Saddlebacks, but settled on Gloucester Old Spots because Karen prefers the lop-eared pigs' temperament. Pigs are easy; you start with a pair of eight-week-old weaners, fatten them up over summer and then take them to slaughter in the autumn.

'If you keep animals for food, the end of their life is as important as the living of it,' says Karen. 'Make sure your local abattoir is one you're happy to use and you can book them in knowing they've had a good life. It's worth planning how you'll load them, and don't feed them on the day. They should be clean and dry. Never take a single animal to slaughter – it'll be less stressful if they have company.'

Very few people I've spoken to make a living from their smallholdings. Karen and Jeff work tirelessly to sell their home-made produce through local farmers' markets. They've held open days, run courses and even supplied their own shops, but as they say, 'Farming is really hard work, but you'll eat top-class food and gain a lot from the satisfaction of a healthy and worthwhile lifestyle.'

Kent College's mixed flock penned and waiting patiently to be drenched by a class of pupils.

LEARNING TO GROW

A young friend of mine, Clio Rudgard-Redsell, wants to be a vet. She passed her GCSE in Environmental and Land-based Science (a qualification no longer available), and she is lucky enough to attend a school with its own farm. According to the School Farms network, there are more than a hundred farms in schools in the UK, with more in the pipeline.

At Chipping Campden School, for example, there are five paddocks where local breeds of Cotswold Longhorn sheep and Gloucester Old Spot pigs are raised and then served in the school canteen; a 1.2ha (3 acre) orchard that pupils have rescued from dereliction, now producing apple juice that's sold in the school and at local farmers' markets; and a flock of hybrid hens which lay eggs that are eaten by the students and staff.

Clio goes to Kent College, a day and boarding school for boys and girls aged 3–18, surrounded by 112ha (280 acres) including 20ha (50 acre) Moat Farm. I visited to see their breeding herd of Dexter cattle, a small flock of sheep, a free-range pig enterprise, a riding school and livery (available to boarding pupils who want to bring their ponies to school with them). The farm is also home to a Noah's Ark collection of hens, ducks, rabbits, guinea pigs and ferrets.

Farmer Palmer gives the students hands-on experience working with the sheep, while others feed and water the poultry and muck out the pens.

A collection of pure-breed ducks wander in from the pond to steal the chicken feed.

The Farmers' Club, an after-school activity, has been going for 63 years and is open to seniors and juniors alike, even those who are fair-weather farmers. Clio told me she joined to spend more time doing things she loved. Her favourite animals are the cows she's known since she was seven, and she will continue to help with them after she has left the school. 'I'm planning to take up veterinary medicine,' she told me. 'I have wanted to do that ever since I was small, but going to the farm has made me more determined and given me valuable work experience and skills I'll need in the future. I've gained a love and understanding of all the animals there.'

The smaller children often start by looking after the guinea pigs and rabbits, learning to clean, feed and care for them. Cuddly, portable and biddable, these animals make an excellent introduction to giving pets a good life. Next in popularity come the hens, a very free-range motley crew of stately buff and black Orpingtons, a few bantams and a handsome Poland cockerel, mysteriously named Janet. The joys of collecting eggs are top on the list of favourite chores, followed by caring for the piglets.

A class appeared while I was there, 20 or so teenage girls and boys dressed in overalls and wellies, divided into groups by teacher Graham Palmer (also known as Farmer Palmer). One set went to drench the mixed flock of Romney, Southdown and Dutch Zwartbles sheep and lambs, steadying the animals between their knees and drenching them for parasites at the back of the tongue with a drenchgun and backpack.

A second group cleaned out the poultry runs, refilled drinkers and topped up feeders, carefully moving a group of pullets from their run into the pell-mell of farmyard life; a third caught and moved the endlessly patient ponies around the paddocks. The fourth group (today's least fortunate) cleaned out the pigsties. All the animals were pelted and spoilt, and the groups worked brilliantly together and had fun.

The highlight of the school year is the annual Kent Show, where the animals are groomed and shown by the students, often winning prizes. As teaching assistant Nicky Manx told me, 'You can always tell the children who've looked after animals – they're the nicest ones.' They're certainly the luckiest.

HOW TO:
MAKE A DUST BATH

Poultry need to dustbathe to keep their plumage clean and it's something they really enjoy. If you are keeping hens in a run you'll need to provide a bathing site for them, and even if they free-range on your plot it's worth offering a purpose-built dust bath to avoid them pockmarking your garden with dips they have scratched out.

Left to their own devices your flock will choose a shady spot, often under a tree or at the back of the border between shrubs where the soil is dry and friable. I recommend you build your henhouse on legs so the hens can shelter and dustbathe underneath – you can see if any vermin are living there too. It's also easier for you to clean out the house if you don't have to bend over too far.

1. First take the measurement between the legs under the house.

2. Cut your timber to fit. I always paint the boards with preservative paint first.

3. Fit them under the henhouse and screw them in place.

4. Fill the dust bath with ash from your woodburning stove, playpit sand or any dry soil from the garden and watch your flock enjoy their bath.

*Which of these ingredients
from The Pig hotel's garden
will feature on today's menu?*

LUXURY ALLOTMENT

**Turning dreams of self-sufficiency into reality sometimes
needs a helping hand. If you're looking for inspiration on how
to transform your garden space into a larder of top-class
ingredients for the kitchen, the place to go is The Pig hotel
near Bath. Here, gardener Ollie Hudson, chef Kamil Oseka
and barman Hywel Day get together on a daily basis to decide
which elements from their garden will be on the day's menu.**

With flocks of quail and hybrid hens to supply eggs, a deer
park for venison and their famous herd of Gloucester Old
Spot pigs, plus extensive kitchen gardens and orchards, the
team at The Pig prove you can grow-your-own with style
and substance. They also have access to local foragers and
small-scale producers within a 40km (25 mile) radius.

The concept of kitchen gardens supplying food for their hotels
isn't new. Gravetye Manor and Le Manoir aux Quat'Saisons
in East Sussex and Oxfordshire respectively have made
formidable reputations doing just that, but this small chain
of five Pig Hotels in the south of England bases its whole ethos
on sourcing ingredients locally and, better still, producing
most of them in the hotels' own gardens.

Gardener Ollie checks a pet pig, while his assistant marks out another row of seeds to be planted.

The 0.6ha (1½ acre) garden was recently rescued from dereliction and bindweed. The neatly regimented raised beds are filled with local, red, sandy soil, enriched with mushroom compost and pig manure, and brassicas, legumes, alliums and roots are rotated. Plants are grown centrally by the hotel chain and delivered as plugs to each garden, where they're grown on until they're ready to harvest. About 20 per cent of the produce is experimental. This is intensive production, but it could easily be scaled down for a family with careful successional planting.

Two decorative, domed 1930s greenhouses are crammed with chillies, Thai basils and stevia – garnishes for the pastry cook. There are currant and berry cages, edged with wild strawberries, and blueberries underplanted with cranberries in special acid soil beds, all yielding up to 1.2kg (2½ pints) of fruit a day in season, and ringed with cold frames to force vegetables. A polytunnel is home to 12 varieties of tomato, of which 'Golden Sweet' is chef's favourite.

Herbs are a big part of this garden's produce. They are supplied, along with cultivation advice, by local herb expert Jekka McVicar, and the beds are crammed with unusual varieties, such as tangerine sage, grey-leaved oyster plant, lovage for salad dressings and creeping savory for salami. Bottles of herb-flavoured oils grace the dining tables, pots of herbs decorate the hotel rooms, and flowers are served as edible plate decorations. Herbs find their way into

A herd of deer roam the deer park and are culled for venison for the hotel's kitchen.

Marigold petals are picked by the boxful to decorate, colour and flavour dishes. Plants are grown in modules and planted out to keep up with the 140 covers that the kitchen supplies each day.

the bar, too: Hywel's team experiment with beech leaves, hyssop and lemon balm in their cocktails and dream up delicious concoctions, such as their cucumber-infused gin with home-made elderflower cordial and a squeeze of lemon.

The menus are determined each day by the produce delivered to the kitchen, and every dish has a garden element. Kamil also smokes home-raised venison and pork products over oak and fruitwoods in his smokehouse, a great way of preserving farmed meats, drawing on his Polish grandmother's recipes where nothing is wasted.

Of Kamil's team of 20 chefs, two are employed just preparing vegetables. They produce 140 covers every day and 250 at weekends, when up to 20kg (44lb) of salad leaves are eaten. The guests seem to appreciate the connection between their meal and its provenance, and usually finish their meal with a wander round the vegetable beds – all part of the fabulous Pig experience.

Beds of colourful country garden flowers and seed heads go into Sheila's pretty wedding bouquets.

BLUE HEN FLOWERS

In 2012, Sheila Hume's son got married. She looked at her fabulously productive 0.2ha (½ acre) garden and thought, 'I can easily grow and arrange the flowers for his wedding.' But the event was in May, and sadly all her flowers bloomed in June. It was then that she decided to look into the business of growing flowers properly.

Luckily, her local farm shop was more than happy to sell the leftover wedding flowers and wanted more. There has always been a shortage of English-grown garden flowers, with 80 per cent flown in from abroad. So Sheila joined Flowers from the Farm, an online network of 200 mostly artisan growers, farmers and market gardeners from Cornwall to Scotland, who supply a range of flowers rarely seen in your local florist's shop or supermarket, and launched her own website.

She decided to concentrate on scented, old-fashioned flowers, such as tulips, narcissi, foxgloves, carnations, sweet peas, dahlias and zinnias. Her website says, 'Whether you're a florist, bride, an enlightened husband looking for a birthday, Christmas or Valentine's present, or just a special bunch of flowers, do get in touch.' Customers can go and pick their own bucket of flowers – good fun, especially for brides and their bridesmaids; or Sheila will decorate a marquee or whole house for a special occasion.

Trying to second-guess exactly when any plant will come into flower in the British climate is impossible, so Sheila invested in a polytunnel which means that she's less at the mercy of the weather's vagaries, and her flowers bloom about a month earlier than they would in the garden. Few extra tools were needed to start her business, although she's now saving up for a small tractor to help move mulch and compost. She has also taken on a trainee to help during the really busy summer months.

Ask any grower what their main enemy is, and slugs usually take first prize. Sheila uses a neighbours' sheep daggings (snipped-off bits of matted wool from the animals' back end) to surround her plants. Wool pellet mulches that will do the same job are available, but Sheila recommends removing and composting them before winter sets in as they offer the perfect cold weather shelter for all sorts of pests.

She grows her flowers in raised beds, using the no-dig system that relies on topping up with compost every year, to create what she calls a 'leaf mould and compost lasagne' in which to grow her annuals and perennials. Seeds (mostly obtained from Mole Seeds) are planted in trays from February onwards, though some (especially *Ammi majus*, nigella and Roger Pearson's sweet peas) are started in the autumn. Foliage plants and herbs are available from the garden proper.

Sheila keeps breeds of hens that lay different coloured eggs. Her Marans lay chocolaty ones, the Leghorns lay white and Araucana crosses lay blue eggs.

The business is named Blue Hen Flowers after the flock of chickens that range about the garden. Eggs of less usual colours appeal to Sheila and her customers, so she keeps Cotswold Legbars (Araucana/ Leghorn crosses) that lay blue and olive eggs and Chocolate Marans and Welsummers that both produce fabulous dark brown ones. Other choices for coloured eggs are Croad Langshans, which produce a plum-coloured egg, and modern hybrids.

Sheila Hume's advice for best flowers to grow for seasonal bunches is:

- Spring: pheasant's-eye narcissus (*N. poeticus* var. *recurvus*) plus 'Greenstar' and 'Mount Tacoma' and *Euphorbia characias* subsp. *wulfenii*.
- Early summer: Sweet peas (*Lathyrus odoratus*) 'High Scent' (also known as 'April in Paris'), 'Just Julia' and 'Lord Nelson'.
- Midsummer: *Digitalis* 'Camelot' series, *Ammi majus*, *Alchemilla mollis*, and clary sage (*Salvia sclarea*) with pineapple mint (*Mentha suaveolens* 'Variegata').
- Autumn: *Dianthus* 'Arabian Night' and 'Green Trick', plus *Nerine bowdenii* with lemon-scented *Pelargonium crispum* 'Variegatum'.
- Winter: *Helleborus* species, *Narcissus* 'Bridal Crown' (grown in the greenhouse) with eucalyptus and artichoke leaves.

Why are beehives usually white? Per-Ola's gaily painted beehives for colour-loving bees.

BEES IN BOHUSLÄN

Per-Ola Hansson lives in a cottage among gardens and allotments in the rocky, humpy Bohuslän landscape near Gothenburg, on the west coast of Sweden. He keeps 20 or so gaily painted hives, each populated by about 50,000 bees, so he is responsible for about a million busy insects. Per-Ola knows a lot about his charges and his enthusiasm is infectious, though not to all of his family – his wife is allergic to bees, but his daughter helps him during harvest time.

He started 12 years ago, working with a mentor and four colonies of Buckfast bees – a calm strain of English bees – and has gradually built up his population, carefully managing his swarms. The bees will forage up to 6km (3¾ miles) from their hives in search of nectar, but here among the carefully tended allotments, there's plenty to keep them occupied. He says, 'Keeping bees is hard work at certain times of the year, but it's relaxing compared to my day job as a train driver.'

Dressed in his white beekeeper's suit – for bees are calmed by the colour white – Per-Ola lights his smoker and puffs smoke into the hives to calm his bees, quietly explaining what he's doing. Lifting each frame, he examines them intently, pointing out the queen, painted with a small dot of nail varnish to distinguish her from her thousands of smaller workers. Hives should be inspected weekly for overcrowding and the larger queen cells destroyed unless the intention is to increase the colony. Varroa mite is a parasitic problem with honey bees the world over, and Per-Ola checks for these pests on the bottom board, treating them with vaporized mineral oil if they are present. He also judges how much honey will need to be harvested.

Collecting honey from the frames in time-honoured fashion to produce jars of golden honey collected from flowers grown for their distinct flavour: blossom, lavender or limeflower.

Per-Ola finds six full frames and carries them into the processing area. Removing the wax from both sides of the frame with a special knife, he hangs them in his gleaming centrifugal extractor. As they start to spin, the honey sticks to the side of the drum, where it oozes down and drips into a bucket. Each frame will produce a glistening 2kg (4½lb) of honey.

As the summer wears on and different flowers bloom, the bees produce honeys with slightly different flavour. In early spring they forage on pussy willow, fruit tree blossom, rosemary and heather; May brings limeflower, and late summer is redolent with lavender. Per-Ola blends the honeys like a winemaker, but the best are sold as single-flower honeys, keenly sought after by honey aficionados.

The season ends early in Sweden with the arrival of colder weather, and the hives are prepared for winter. The entrances are narrowed with chamfered blocks of wood and protected with a strip of metal to deter mice, which would otherwise chew through to steal the honey once the bees are quiet. Until spring comes and nectar is available again, Per-Ola's bees are given a winter's supply of sugar syrup spread on straw and he leaves them undisturbed until the spring, adding a layer of blanket when the temperature falls. He never opens the hive when it's really cold, since bees become cold very quickly.

During the winter Per-Ola cleans and overhauls equipment and orders seeds and plants to tempt his charges next year. Herbs such as hyssop, thyme and borage are particularly beneficial.

If you keep bees and need to boost the vitality of your vegetable plot, sow a green manure such as alfalfa, phacelia or clover – the bees will love them. Your neighbours will appreciate your bees as they will pollinate plants in their gardens. Encourage them to lay aside insecticide sprays and plant single native flowers that will feed the bees throughout their season.

HOW TO:
BUILD A BEE HOTEL

Pollinating insects are vital to crops, so we need to look after them. Make sure your garden offers year-round food, water and habitat where bees can overwinter and nest. Leave piles of logs and stones, and don't tidy your garden until winter has passed. This hotel will make a desirable residence for bees and hoverflies; there are more than 240 species of wild bees that don't live in hives.

Use up bits of leftover timber (preferably not wood that has been recently treated with preservative). You'll need 700mm (28in) of 75 x 22mm (3 x $^{7}/_{8}$in) cut in two, and 330mm (13in) of 75 x 38mm (3 x 1½in) cut into three pieces, plus 12 50mm (2in) screws and lengths of hollow stem for the nests. Be careful when harvesting stems not to use giant hogweed as this will burn your skin; try bramble, common hogweed, reed or bamboo.

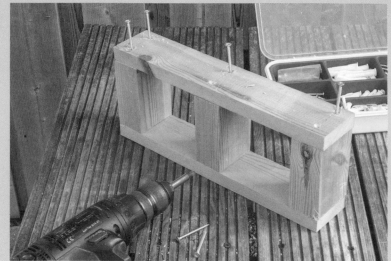

1. Saw some timber offcuts to size (see p.120), then screw the smaller pieces on either end and in the middle of the longer piece.

2. Screw on the top piece.

3. Drill extra holes in the wood for smaller insects.

4. Cut the sections of stem to fit and glue them inside the two cavities.

5. Hang your bee hotel against a fence or wall in a sunny (south-facing postion if possible), sheltered spot about 30cm (12in) above ground level.

MUCK INTO MAGIC

Whatever you grow and wherever you grow it, to get the best from your plot, there are certain basic tenets that apply. Your soil needs to be enriched, seeds must be sown, pollinating insects should be encouraged, and waste needs to be effectively managed.

Of course you can buy soil enhancers, compost and mulches; you can spend a fortune on commercial seeds, buy in pollinating insects and pay to have waste removed; but why not save precious money? Smallholders have been composting, seed-saving, spreading soil improvers and helping wildlife for generations, and the satisfaction of following time-honoured practices is part of the joy of raising produce and growing plants.

All the smallholders we visited have turned muck into magic, sown seeds of success and worked together to fulfil their ultimate goal – to produce food humanely and in an ecologically benign way.

STARTING FROM SCRATCH

At first glance, Charlotte and Donald Molesworth's garden looks as though it has been in the family for generations, probably the result of sustained work by a whole team of gardeners and certainly the product of several hard-earned fortunes. Of course, the Molesworths have put their hearts and backs into their garden, but it's their imagination and ingenuity that have given birth to its soul.

Creating a garden like this on a shoestring is an art – but combine Donald's background in farming, Charlotte's career as an artist and their shared love of the natural world, and you have exactly the right ingredients for the ideal garden. They found the old kitchen garden site in 1983, complete with a dilapidated bothy, piggery and pottery, and rescued it from dereliction.

Bringing life to their garden on a budget over the past 30 years has been a lesson in thrift. They still go to farm sales and reclamation yards but admit it's harder to find good pieces nowadays, and suggest keeping an eye open for building and demolition sites to search out useful agricultural and industrial pieces.

Charlotte and Donald have created a garden packed full of great ideas and money-saving techniques among the fabulous topiary fantasies.

Their famous hedges and topiary were all planted from seedlings or cuttings; paths have been paved with reclaimed bricks that are 'over-fired seconds'; gates were rescued from skips and fences made with home-grown hazel. With an imagination like Charlotte's, a redundant tennis court soon becomes a fruit cage, plastic barrels are transformed into water butts and a metal pigswill boiler is turned into an eye-catching plant container.

The Molesworths say, 'Going shopping isn't our first reaction to fulfilling a need. People pass things on.' A list of most gardeners' regular outgoings includes seeds, compost and plants. This couple's home-composting regime includes layering all their old clothes, paper and cardboard along with kitchen, garden and animal waste (the last from their elderly Jacob, Shetland and Soay sheep or rescued battery poultry). And to them, do-it-yourself propagation is what gardening is all about.

Charlotte Molesworth selects the best plants to save seed from and then collects them when dry, before the autumn weather sets in.

The Molesworths' beehives are full of worker bees who make sure flowers are pollinated so there's seed to be stored carefully in lined, damp-proof drawers, safe from predators.

- To propagate your own plants, collect the seed of interesting non-hybrid varieties on a dry sunny day after the dew has dried and pop into labelled envelopes.
- 'Dry' seeds such as those of pulses, peppers, onions and most herbs and flowers should be stored in an airy place until their pods or husks are completely dry. Then crumble the pods or husks and winnow the seeds to separate them from the chaff by placing them in a bowl and swirling gently around. The seed will sink to the bottom and the chaff can be carefully removed.
- Cucumber and aubergine seeds need to be collected from the pulp of their fruit. Scoop the seeds and pulp into a bowl and add water. The seeds will sink. Rinse them in a sieve and leave to dry on a shiny plate in an airy place. Store your seed in jars.
- Some seeds, including tomatoes, melons, squash and cucumbers, must be fermented to remove germination-inhibiting coatings. Put the seeds and pulp into a jam jar, cover with water and leave in a warm place until a layer of bubbles forms on the surface. Drain then clean as above.
- Store your seed until planting time, making sure that rodents can't get at it. Dusting with a little diatomaceous earth will prevent insect infestation. Special seeds can be swapped with friends on garden visits and make great presents.

Green & Serene
fine homegrown
plants

Pesticide Free

S. Henderson . 07738 515 263
GreenandSerene@hotmail.co.uk

WASTE TO WONDER

How to recycle waste effectively is one of the smallholder's biggest quandaries. Turning animal and vegetable by-products into something useful and enriching to condition our soil, without polluting our surroundings or expecting local authorities to shoulder the burden, requires thought and planning.

Serena and Marcus Henderson have the process down to a fine art. They rent small pieces of ancient orchard from local farmers and turn the apples into pure gold – artisan cider. The fruit would otherwise have gone to waste as growers now plant small dwarf trees that are easily managed, but some retain an affection for these magical spots with their standard trees and heritage varieties and leave them to wildlife.

The Hendersons manage these small pockets of horticultural history kindly and make a living selling their cider online and at local fairs, markets and festivals. 'It started as fun, but now we can't keep up with demand,' says Serena. They believe in localism and source heritage varieties which they juice in traditional oak presses, leaving some to mature in old barrels and mulling others.

Unwanted billy goats, hens and turkeys all find a welcoming home on Serena and Marcus' smallholding.

Most of the pommace, or waste residue, is taken away by a local shepherd to feed his flock over winter. Some is kept and added to their massive compost heap in among the fruit trees, alongside manure and bedding from their motley flock of rescued billy goats, much-loved hens and a fabulous Bourbon Red stag turkey, Barney, who'll never have to worry about Christmas. They pile on kitchen and garden waste and turn the compost regularly with a digger basket on an old farm tractor, and when it's ready Serena grows a fabulous crop of pumpkins on this glorious compost mound. She pops seeds into large pots and plants them out when the seedlings are established. They're robust, pest-free and great fun to grow.

I always plant a few squashes in my own small compost heap and they ramble across the dead hedge behind my garden. The young shoots can be eaten steamed as a side dish, and the pumpkins themselves are used in savoury or sweet dishes, made into jam, or, in the hands of experts, even turned into cider.

Apples are grown for their cider and sold at festivals and fairs all over the county, and the waste products or pommace is composted with animal waste.

Most soils can be improved by adding organic manure; it adds nourishment and substance to sandy soils and lightens clay ones. Animal excrement, added to their shed hair and feathers, plus litter or bedding – especially straw, ribbed paper or card and hemp – is the best source of garden manure, and best left to rot under cover. Wood shavings and sawdust are often treated with preservatives and take too long to rot down, so should be avoided.

- Horse and cow manure are the most valuable, especially as a soil conditioner, but should be kept under cover as they lose value when wet. Goat and sheep droppings are richer in nitrogen. Pig, cat and dog faeces should be avoided for disease reasons. Horse manure will always contain weed seeds.
- Poultry manure is highest in nitrogen and phosphorus and the nutrients are quickly available to plants. It should be stored in a dry place, and manure from intensively farmed birds should be avoided. Leave to rot because fresh chicken manure from flocks fed on pellets can burn plants, then add to the soil at a rate of 600g per 1sq. m (1lb per 1sq. yd). The droppings of birds raised outdoors may contain weed seeds.
- Seaweed is nearly as beneficial as farmyard manure, although it is lower in phosphates but richer in potash.
- Always leave any manure to rot down among layers of bedding and general garden waste. Never apply it fresh.

HOW TO:
MAKE NETTLE FERTILIZER

While nettles are generally regarded as unwelcome weeds, they make a nutritious plant fertilizer. Wearing gloves, pick your stinging nettles young, as they'll break down more quickly. Scrunch them up by hand or run a lawn mower over them. Pop them in a bucket, wedge them down with a brick or a flowerpot and then cover with rainwater. Site the bucket at the bottom of the garden, as it will get quite smelly during the 3–4 weeks it takes the fertilizer to brew. Strain and then dilute with water, one part nettle juice to ten parts water – it will look like weak tea. Use this solution to water around the base of any vegetable plants and they'll benefit from the nettles' nitrogen and potassium as well as other crucial minerals. Alternatively, make a foliar feed by diluting the brew to 1:20 and spray on the leaves.

1. Wearing gloves, pick a bunch of young nettles carefully, avoiding the roots.

2. Bruise the leaves and stems by twisting. Place in a bucket, weighed down by a brick.

3. Steep by covering with water and leave for 3–4 weeks in a place where no one will be subjected to the smell.

4. Water plants with liquid diluted in water 1:10 or spray as a foliar feed diluted 1:20.

FOREST GARDENING

My son Jacques is new to gardening and has started with a difficult shady plot under giant oak trees. His plan is to grow edibles under the trees using forest gardening techniques, where plants are grown in layers based on woodland ecosystems, incorporating fruit and nut trees, herbs, vines and perennial vegetables. He has already planted local varieties of cobnut trees, and elder bushes that will grow under their canopy.

This a low-maintenance way of growing food, though it takes a bit of effort to start with. You work with nature by mimicking principles of a natural forest. Many cultures have a history of growing food this way; only western agriculture is based on monocultural food production. To make maximum use of space, ideally seven layers of planting are used:

- Canopy: trees and shrubs – fruit and nut trees.
- Understorey: low-growing, shade-tolerant trees – dwarf fruit and nut trees.
- Shrub layer: woody plants, such as raspberries, currants and *Elaeagnus umbellata*.
- Herbaceous layer: herbs and perennial vegetables – Welsh onions and artichokes.
- Vertical layer: climbing plants and vines, such as Japanese wineberries and kiwis.
- Ground layer: creepers such as alpine strawberries
- Rhizosphere: root vegetables.

No space is wasted: pumpkins are planted in the compost heap to romp along the fence.

Once established, this type of garden needs no watering, digging or feeding. The plants provide their own nutrient requirements through annual leaf fall and from mineral accumulators, for example comfrey, and nitrogen-fixing plants such as elaeagnus and clovers. It's also a good place for children to play and wildlife to flourish. To start with, though, Jacques needs to improve the clay soil structure, left compacted by builders, by incorporating leaf mould and compost. Since he's keen to recycle kitchen and garden waste, he has developed an enviable system.

Next to the fence at the bottom of the garden, he has set up a 'dead hedge' to store branches too big to compost and too small to burn in his woodburning stove, where prickly and awkward branches will quietly decompose over the years. It also makes an effective windbreak and habitat for insects, especially beetles, and birds.

In front of the hedge are three compost bays made of wooden pallets where garden waste slowly composts down. Using layers of green and brown garden waste (see below) and chicken manure from my hens, Jacques produces a crumbly mulch that he layers on top of the existing earth, using a no-dig system, where insects and worms help to improve the soil. He also mulches his holding beds, where he's temporarily growing berry bushes and rhubarb to plant out when the soil is ready.

Leaves are stored in large open-weave builders' bags to make leaf mould for soil conditioner. Oak leaves take over a year to rot down, but ash and elder are faster. Running a mower over them on paths and lawns helps to break them down.

Jacques' young family produces massive quantities of food waste, so he has invested in a 200 litre (44 gallon) insulated Hotbin composter (the size of a wheelie-bin) that super-heats waste to 60°C (140°F) and produces mature compost in just 90 days. By combining layers of waste with bulking agent (shredded brown garden waste) and shredded paper and card, he can cut down on his household waste that goes to the local council and help to enrich his forest garden.

GREEN WASTE
High in nitrogen: animal manure, annuals and perennials, veg and fruit peelings and greenery, tea leaves and coffee grounds, grass mowings and soft hedge clippings.

BROWN WASTE
High in carbon: shredded cardboard and paper, cotton wool, eggshells and boxes, hair, straw, wood ash, twigs, sawdust and stems.

These leek flowers left to go to seed are magnets to bumblebees.

BEFRIENDING THE POLLINATORS

The Abbey Physic Community Garden in Kent is a charity set up to provide a friendly working environment and support for adults with mental health issues or those who are socially isolated. However, anyone can enjoy its tranquil ancient walled garden, volunteers are welcome, and this organic garden has won awards for its efforts to befriend all members of its local community, including beasts, bugs and birds.

People often think that a wildlife-friendly plot is an untidy plot. Not so: virtually any smallholding can be turned into a haven for insects, birds and other species by providing just three elements – water, food and shelter. Alongside a working vegetable plot at the Abbey Physic Community Garden, fruit trees, blossom, and flowering herbs scent the air; it's a paradise that proves that a few changes in the way you manage your garden, alongside gardening organically rather than spraying insecticide, can bring major benefits to the creatures that have made it their home.

If you want to grow fruit and vegetables on any scale, you need pollinators to fertilize your plants to produce fruit and seeds. The past half century has seen a decline in the number of bees and other beneficial insects that visit our plants to feed on pollen and nectar, especially bumblebees (so useful to pollinate early fruit blossom), solitary bees, hoverflies, butterflies, moths and pollen beetles.

It's not just year-round food plants that insects need, but habitat. Build your own bee hotel and leave neat piles of stones and wood.

Keeping honey bees may be one answer to the problem of fruit pollination, but encouraging insects generally is another option. The Abbey Physic Garden has chosen to grow a wide variety of flowers to supply pollen and nectar throughout the year, starting with mahonia, wild cherry and pulmonaria right through to late-flowering ivy. Lists of wildlife-friendly plants are available on many websites. Trees, climbers and shrubs are also invaluable to insects, especially in urban areas, like this one.

The single easiest way to encourage wildlife into your garden is to build a pond, since insects also need water. The Abbey has a small pond, not just to encourage amphibians that will eat those insects you don't want to encourage, but with a shallow, sloping side that any animal can access and native plant cover where insects can stop and have a drink. Compost heaps will attract slow worms and grass snakes that'll eat your garden's pests.

Water is essential for insects as well as amphibians, birds and mammals.

Leaving a pile of dead wood in a corner and another of stones provides habitat and nest sites for other insects, and the Abbey has a range of bug and beetle hotels in the form of piled pallets with all their recesses stuffed with recycled bits and pieces such as hollow stems, stacked tiles, cut bamboo, pierced bricks, card tubes, straw and small pots. A dead hedge made of decaying twigs too big to compost and too small to burn also makes a desirable home or nest.

A small patch of meadow grass will offer food and shelter to insects, and if you leave this and your garden plants uncut overwinter, you'll provide useful hibernating habitat. Let daisies and dandelions grow in your lawn, like they do at the Abbey Physic Garden, and help those helpful bees and bugs to find nectar.

- Mason bees like small rock piles, or holes in wood 15cm (6in) deep.
- Leaf-cutter bees nest in holes in dead wood or old walls.
- Compost heaps offer nest sites for hoverflies, whose larvae feed on rotting organic matter.
- Solitary bees make individual nests for their offspring in hollow stems or holes in wood in a warm sunny spot.
- Lacewings and ladybirds will eat aphids. They like to live in rolled corrugated cardboard or bundles of hollow sticks.
- Bumblebees will nest in an upturned flowerpot with access through the draining hole. Mine live in the ventilation bricks around the base of my house.

Farming in groups brings economies of scale, as well as good company.

THREE GO SMALLHOLDING

Several winters ago photographer Bill Mason and his neighbours Phil and John were sitting at his kitchen table discussing the joys of growing and raising their own food. All three gardened and kept a few animals, but wondered how much more productive they would be if they pooled their land and labour and worked a smallholding together.

After meeting frequently in the local pub to make plans, they decided to farm Bill's 4ha (10 acre) semi-derelict orchard/nuttery and to produce the food they most liked to eat and the provenance of which they preferred to take into their own hands. They plumped on pigs for sausages, geese for special occasions, and fruit, nuts and vegetables.

So they opened a joint bank account and invested equal down payments for stock, fencing and feed, then made a plan to divide the labour depending on specialisms: Bill was in charge of veg and the existing fruit and nut trees; Phil already managed a flock of geese, and John liked the idea of learning how to keep pigs and make charcuterie. Of course they would help each other out when needed, and work together to set up.

There is legislation governing the keeping of stock, and dealing with this was one of the first tasks the trio had to undertake. You have to register flocks of poultry of 50 birds or more in the UK, and if you keep pigs, cattle, sheep or goats you'll need identification numbers; see defra.gov.uk for details. Every country has different regulations that need to be strictly adhered to, in order to keep, move or sell livestock.

Next, a concrete hard-standing was laid on the orchard's edge near the standpipe to keep mud at bay, with space for two straw-filled pig arks and food troughs. The plot next to it was stock-fenced off with four-strand electric fencing and designated as the vegetable garden to be set up next year after the porkers had cleared and manured the land.

Looking after stock is much easier if the work of feeding, watering and mucking out is shared.

Then Phil's existing flock of geese was corralled into a corner of the orchard next to the pond and a wooden shelter filled with dry straw bedding, surrounded by netted fencing, was built with a wide doorway (unlike most types of poultry, geese need more than a pophole). The flock would spend their days grazing grass in the orchard and be locked away at night. Geese are hardy creatures: house them for protection from predators rather than night-time comfort.

Geese need grass, grain, grit and water to survive. They are herbivores (rather than omnivores like most other poultry) requiring large amounts of good rotated grazing, supplemented with morning and evening helpings of 80 per cent mixed corn and 20 per cent pellets (top up the pellets if you want eggs). Keep all feed in vermin-proof metal containers.

A flock of geese works well in a mixed smallholding, grazing on much shorter grass than other stock. They will flourish in an orchard, finishing up windfalls. They don't thrive on rough pasture, so when grass loses its goodness in winter, substitute leafy veg and extra corn or pellets. Be warned that geese mess a lot, so don't encourage back-door snacking.

Clean drinking water is essential for waterfowl and a large pond is necessary for their preening and general well-being. Geese lay a clutch of eggs in the spring (though some breeds, such as Chinese Browns, may lay up to 60 or so), the eggs are excellent for baking. You don't need a gander for your goose to lay, but if you keep one and the eggs are fertile, you can let their mum hatch your own flock for next year.

Phil buys in eggs in spring, hatches them in incubators, raises the goslings under heat, feeds them up during the summer and autumn, then takes them to slaughter at ten months. Before slaughter, birds of any kind should be starved for 12 hours, but allowed access to water.

Costly farm machinery, stock and animal housing becomes a possibility if expenses and maintenance are shared between friends.

Choosing a local slaughterhouse is one of the most important steps to take when considering keeping animals for meat, and should be one of the main criteria when buying land for this purpose. Home slaughter is possible for geese and the Humane Association website gives excellent advice, but always go on a course first, and learn how to do it properly.

John went on a course to learn pig keeping and how to make charcuterie. The group decided to buy six, easy-to-keep Saddleback growers (mixed gilts and boars) from a local pig breeder, though farming directory websites offer a large variety of breeds. Growers are ten weeks old, well weaned and stronger than weaners, which are just 6–8 weeks.

DEFRA was contacted, holding and herd numbers were obtained and the animals were tagged. Throughout the summer the herd was rotated in electric-fenced areas that needed to be cleared. Water was provided in a barrel nipple drinker or hog watering system placed on a sheet of plywood, and they were fed on grower then finisher pignuts, plus leftover veg and windfalls. Pigs can be fed waste from the garden, but not from the kitchen.

No one wanted to see them go, but come autumn the ground was turning into a swamp. John put the trailer into their paddock and fed them in it for a few days so they'd get used to it, and off they went to the slaughterhouse. He bought the carcasses back to butcher himself, because the meat was for home consumption. Bill, Phil and John and their families enjoyed their first year's produce, stored in extra freezers, and made plans for the future.

Geese don't need more than a few inches of water to preen and dip their heads, but you shouldn't deny them the pleasure of open water.

HOW TO:
BUILD A DEAD HEDGE

A dead hedge is a great way to slowly compost wood that's too thick to compost in a bin or heap and too thin to burn effectively. It also makes a good, natural-looking barrier between parts of the garden and is the perfect habitat for insects and mammals, especially once the bottom pieces start to rot. If you add hollow stems, pollinating insects may nest and hibernate inside them.

This one is sited between a fence and the backs of compost heaps. The space is particularly useful when renovating a garden, as it provides a home for lots of debris and saves a fortune in refuse disposal. It's a good way to screen off areas of the garden that you'd rather hide, or to provide a safe fence to make a pond less dangerous.

1. Plant preserved wooden posts at 1m (3¼ft) intervals in two rows, leaving a 50cm (20in) gap in the middle.

3. Continue to build up the dead wood to the top.

2. Weave in stems and sticks in layers between the posts.

4. The bottom layers will eventually rot down, leaving space to top up.

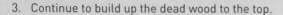

RESOURCES

ALLOTMENTS

communitygarden.org – Community gardens in the US.
gardenorganic.org.uk – Long-established charity leading the movement for an organic future.
nsalg.org.uk – The National Allotment Society.
soilassociation.org – Organic certification body.

ANIMAL WELFARE

americanhumane.org – The American Humane Society.
aspca.org – The American Society for Prevention of Cruelty to Animals.
hsa.org.uk – The Humane Slaughter Association for the treatment of food animals.
rspca.org.uk – The Royal Society for the Prevention of Cruelty to Animals.

BEES

abfnet.org – The American Beekeeping Federation.
bbka.org.uk – The UK's leading beekeepers' association.
losangelescountybeekeepers.com – Very informative beekeeping website.
beeurban.se – Swedish beekeepers
urbanbees.co.uk – Support for urban beekeepers.

COMMUNITY GARDENING

rhs.org.uk/communities – Find a community garden near you.
cityfarmer.info – Urban farming news.

COMPOST

hotbincomposting.com – For fast composting in HotBins.
planetnatural.com – Information about composting plus organic garden supplies in the US.

COURSES

cityandguilds.com – Agricultural and horticultural courses.
farmskills.co.uk – Practical commercial farm-based training.
rhs.org.uk – The RHS offers training in growing fruit, veg and herbs at 80 centres.

FLOWERS

bluehenflowers.co.uk – For locally grown garden flowers.
flowersfromthefarm.co.uk – Find a grower near you.
slowflowers.com – Directory of eco-conscious American flower growers.

GOATS

modernfarmer.com – Backyard goat keeping in US.
pygmygoat.co.uk – For info about pygmy goats.
Ukgoats.co.uk – All you need to know about goats.
Theselfsufficientliving.com – A comprehensive US site with info on goats.

HERBS

herbsociety.org.uk – Charity dedicated to encouraging use of herbs.
herbsociety.org – The Herb Society of America.

MANURES

gardeningdata.co.uk – Information, advice and links to suppliers.
greenmanure.co.uk – Green manure seeds.

NUTS

Profitableplants.com – Growing nut trees in the US.
walnuttrees.co.uk – The Walnut Tree Company for nut trees.

ORCHARDS

theurbanorchardproject.org – Find a community orchard in the UK.
Common Ground – Local projects with orchards and woodlands.
Ptes.org/orchardmaps – A charity mapping orchards in England and Wales and monitoring wildlife.
rhs.org.uk – Advice on growing fruit.

OUTDOOR EATING

anevay.co.uk – Next-generation woodburning stoves.
buckstove.com – Fire pits and outdoor stoves in the US.
wildstoves.co.uk – For a good range of outdoor stoves.

PERMACULTURE

earth-ways.co.uk – Information on how to create your own permaculture design.
permaculture.org.uk – The Permaculture Association.

PIGS

backyardpigs.com – Advice on pig-keeping and good recipes for pig meat products.
britishpigs.org.uk – The website of the British Pig Association, with information on breeds, breeders, shows and much more.

POLLINATION

bumblebeeconservation.org – How to provide the right habitat for bumblebees.
buzzaboutbees.net – Information about bees, including how to encourage them into your garden.
wildlifetrusts.org.uk – Garden wildlife advice.

POULTRY

agric.wa.au/poultry-birds – Poultry keeping in Australia.
beginningfarmers.org – Small-scale poultry farming in the US.
callducks.net – The Call Duck Association UK.
bhwt.org.uk – The British Hen Welfare Trust, for a free-range future.
flyte-so-fancy.co.uk – Housing and equipment.
henkeepersassociation.co.uk – Advice for those who keep poultry for pleasure.
metzerfarms.com – Hatchery in the US with practical advice on the website.
Omlet.co.uk – Modern and traditional housing.
poultryclub.org –The Poultry Club of Great Britain, safeguarding the interests of pure breeds.
poultrykeeper.com – For general information.
waterfowl.org.uk – The British Waterfowl Association.

PRESERVING

nchfp.uga.edu – Advice on most methods of home food preservation.
lakeland.co.uk – For cooking utensils and preserving jars.

RARE BREEDS

livestockconservancy.org – Rare breeds conservancy in the US.
rbst.org.uk – The Rare Breed Survival Trust.
rbta.org – The Rare Breeds Trust of Australia.

REGULATIONS

defra.gov.uk – Essential information about registration and identification in the UK.
fda.gov/Food/Guidance – Standards for growing produce in US.
food.gov.uk – Information about food safety and hygiene.
nfsco.co.uk – The National Fallen Stock Company.
usda.gov/organic-agriculture.html – Organic food regulations in the US.

SCHOOLS

farmgarden.org.uk/school-farms-network – Information on schools that teach farming.
kentcollege.com – Independent co-educational school that teaches agriculture.

SEEDS

plant-world-seeds.com – Growers of more than 3,000 varieties of flower, herb and veg seed, some rare and exotic.
realseeds.co.uk – Heritage vegetable seeds for the home gardener.

SHEEP

nationalsheep.org.uk – Advice on keeping sheep.
samphireshop.co.uk – Keep up to date with Karen Nethercott.
Motherearthnews.com – Small-scale sheep farming in the US.

SMALLHOLDING

accidentalsmallholder.net – Plenty of advice on the website and details of smallholder courses in Scotland.
countryfarm-lifestyles.com – Self-sufficiency and small-scale farming in the US.
countrysidenetwork.com – Guides to keeping poultry, pigs and rabbits plus a list of suppliers in the US.
Hobbyfarms.com – For info on nut trees in the US.
hobbydierhouder.nl – Dutch Smallholders Association
molevalleyfarmers.com – Smallholding and farming equipment.
smallplotbigideas.co.uk/reference-information/smallholding societies/ – Links to British smallholding groups.

URBAN GARDENING

farmgarden.org.uk – Federation of City Farms and Community Gardens.
globalgeneration.co.uk – Skip garden.
urbanfarming.org – Urban farming in the US.

WINE

ukva.org.uk – The United Kingdom Vineyard Association.

WORKING HOLIDAYS

Workaway.org – For working holidays all over the world.

CONTRIBUTORS' WEBSITES

abbeyphysiccommunitygarden.org
bluehenflowers.com
brooklynfarmgirl.com
davenportvineyards.co.uk
globalgeration.org.uk
heallocal.com
hencorner.com
kentcider.co.uk
kentcollege.com
maddocksfarmorganics.co.uk
royalmail.com.au
samphireshop.co.uk
thepighotel.com
thepottingshedholidaylet.com

AUTHORS

Photograph by Charlie Colmer

Photograph by Annalisa Mason

WORDS BY
FRANCINE RAYMOND

Having a productive garden is one of Francine's greatest pleasures. She writes about her experiences for *The Sunday Telegraph* and *Gardens Illustrated* and blogs at kitchen-garden-hens.co.uk. After a lifetime on an acre in Suffolk populated with hens and ducks, she now gardens a small town plot by the sea in Whitstable with the help of her grandsons and a few bantams.

FRANCINE'S ACKNOWLEDGEMENTS

Thank you so much for letting us into your gardens, smallholdings and lives – without your stories, journeys and experiences this book wouldn't exist; without the expertise of all at Pavilion, they would all remain a pipe dream and without Bill Mason's images turning them into reality, we wouldn't have a book.

I hope readers will be inspired to try their hand at growing and raising produce, remembering that food raised with love and care will taste better and enrich the lives of all who enjoy it.

PHOTOGRAPHY BY
BILL MASON

Photojournalist and photographer Bill Mason enjoys meeting food producers and artisans and documenting their skills. Inspired by their lifestyles, he now runs a smallholding at his farm in Kent where he keeps pigs and grows fruit and vegetables for his family. He shares his time travelling between the UK and Sweden.

OTHER PICTURE CREDITS

pp 24–27: Pamela Reed; **pp 42–45**: Royal Mail Hotel; and **pp 52–55**: Mockingbird Meadows.